# *p*-adic functional analysis

## PURE AND APPLIED MATHEMATICS

**A Program of Monographs, Textbooks, and Lecture Notes**

## LECTURE NOTES IN PURE AND APPLIED MATHEMATICS

1. *N. Jacobson,* Exceptional Lie Algebras
2. *L.-Å. Lindahl and F. Poulsen,* Thin Sets in Harmonic Analysis
3. *I. Satake,* Classification Theory of Semi-Simple Algebraic Groups
4. *F. Hirzebruch, W. D. Newmann, and S. S. Koh, Differentiable Manifolds and Quadratic Forms*
5. *I. Chavel,* Riemannian Symmetric Spaces of Rank One
6. *R. B. Burckel,* Characterization of C(X) Among Its Subalgebras
7. *B. R. McDonald, A. R. Magid, and K. C. Smith,* Ring Theory: Proceedings of the Oklahoma Conference
8. *Y.-T. Siu,* Techniques of Extension on Analytic Objects
9. *S. R. Caradus, W. E. Pfaffenberger, and B. Yood,* Calkin Algebras and Algebras of Operators on Banach Spaces
10. *E. O. Roxin, P.-T. Liu, and R. L. Sternberg,* Differential Games and Control Theory
11. *M. Orzech and C. Small,* The Brauer Group of Commutative Rings
12. *S. Thomier,* Topology and Its Applications
13. *J. M. Lopez and K. A. Ross,* Sidon Sets
14. *W. W. Comfort and S. Negrepontis,* Continuous Pseudometrics
15. *K. McKennon and J. M. Robertson,* Locally Convex Spaces
16. *M. Carmeli and S. Malin,* Representations of the Rotation and Lorentz Groups: An Introduction
17. *G. B. Seligman,* Rational Methods in Lie Algebras
18. *D. G. de Figueiredo,* Functional Analysis: Proceedings of the Brazilian Mathematical Society Symposium
19. *L. Cesari, R. Kannan, and J. D. Schuur,* Nonlinear Functional Analysis and Differential Equations: Proceedings of the Michigan State University Conference
20. *J. J. Schäffer,* Geometry of Spheres in Normed Spaces
21. *K. Yano and M. Kon,* Anti-Invariant Submanifolds
22. *W. V. Vasconcelos,* The Rings of Dimension Two
23. *R. E. Chandler,* Hausdorff Compactifications
24. *S. P. Franklin and B. V. S. Thomas,* Topology: Proceedings of the Memphis State University Conference
25. *S. K. Jain,* Ring Theory: Proceedings of the Ohio University Conference
26. *B. R. McDonald and R. A. Morris,* Ring Theory II: Proceedings of the Second Oklahoma Conference
27. *R. B. Mura and A. Rhemtulla,* Orderable Groups
28. *J. R. Graef,* Stability of Dynamical Systems: Theory and Applications
29. *H.-C. Wang,* Homogeneous Branch Algebras
30. *E. O. Roxin, P.-T. Liu, and R. L. Sternberg,* Differential Games and Control Theory II
31. *R. D. Porter,* Introduction to Fibre Bundles
32. *M. Altman,* Contractors and Contractor Directions Theory and Applications
33. *J. S. Golan,* Decomposition and Dimension in Module Categories
34. *G. Fairweather,* Finite Element Galerkin Methods for Differential Equations
35. *J. D. Sally,* Numbers of Generators of Ideals in Local Rings
36. *S. S. Miller,* Complex Analysis: Proceedings of the S.U.N.Y. Brockport Conference
37. *R. Gordon,* Representation Theory of Algebras: Proceedings of the Philadelphia Conference
38. *M. Goto and F. D. Grosshans,* Semisimple Lie Algebras
39. *A. I. Arruda, N. C. A. da Costa, and R. Chuaqui,* Mathematical Logic: Proceedings of the First Brazilian Conference
40. *F. Van Oystaeyen,* Ring Theory: Proceedings of the 1977 Antwerp Conference
41. *F. Van Oystaeyen and A. Verschoren,* Reflectors and Localization: Application to Sheaf Theory
42. *M. Satyanarayana,* Positively Ordered Semigroups
43. *D. L Russell,* Mathematics of Finite-Dimensional Control Systems
44. *P.-T. Liu and E. Roxin,* Differential Games and Control Theory III: Proceedings of the Third Kingston Conference, Part A
45. *A. Geramita and J. Seberry,* Orthogonal Designs: Quadratic Forms and Hadamard Matrices
46. *J. Cigler, V. Losert, and P. Michor,* Banach Modules and Functors on Categories of Banach Spaces

47. *P.-T. Liu and J. G. Sutinen,* Control Theory in Mathematical Economics: Proceedings of the Third Kingston Conference, Part B
48. *C. Byrnes,* Partial Differential Equations and Geometry
49. *G. Klambauer,* Problems and Propositions in Analysis
50. *J. Knopfmacher,* Analytic Arithmetic of Algebraic Function Fields
51. *F. Van Oystaeyen,* Ring Theory: Proceedings of the 1978 Antwerp Conference
52. *B. Kadem,* Binary Time Series
53. *J. Barros-Neto and R. A. Artino,* Hypoelliptic Boundary-Value Problems
54. *R. L. Sternberg, A. J. Kalinowski, and J. S. Papadakis,* Nonlinear Partial Differential Equations in Engineering and Applied Science
55. *B. R. McDonald,* Ring Theory and Algebra III: Proceedngs of the Third Oklahoma Conference
56. *J. S. Golan,* Structure Sheaves Over a Noncommutative Ring
57. *T. V. Narayana, J. G. Williams, and R. M. Mathsen,* Combinatorics, Representation Theory and Statistical Methods in Groups: YOUNG DAY Proceedings
58. *T. A. Burton,* Modeling and Differential Equations in Biology
59. *K. H. Kim and F. W. Roush,* Introduction to Mathematical Consensus Theory
60. *J. Banas and K. Goebel,* Measures of Noncompactness in Banach Spaces
61. *O. A. Nielson,* Direct Integral Theory
62. *J. E. Smith, G. O. Kenny, and R. N. Ball,* Ordered Groups: Proceedings of the Boise State Conference
63. *J. Cronin,* Mathematics of Cell Electrophysiology
64. *J. W. Brewer,* Power Series Over Commutative Rings
65. *P. K. Kamthan and M. Gupta,* Sequence Spaces and Series
66. *T. G. McLaughlin,* Regressive Sets and the Theory of Isols
67. *T. L. Herdman, S. M. Rankin III, and H. W. Stech,* Integral and Functional Differential Equations
68. *R. Draper,* Commutative Algebra: Analytic Methods
69. *W. G. McKay and J. Patera,* Tables of Dimensions, Indices, and Branching Rules for Representations of Simple Lie Algebras
70. *R. L. Devaney and Z. H. Nitecki,* Classical Mechanics and Dynamical Systems
71. *J. Van Geel,* Places and Valuations in Noncommutative Ring Theory
72. *C. Faith,* Injective Modules and Injective Quotient Rings
73. *A. Fiacco,* Mathematical Programming with Data Perturbations I
74. *P. Schultz, C. Praeger, and R. Sullivan,* Algebraic Structures and Applications: Proceedings of the First Western Australian Conference on Algebra
75. *L Bican, T. Kepka, and P. Nemec,* Rings, Modules, and Preradicals
76. *D. C. Kay and M. Breen,* Convexity and Related Combinatorial Geometry: Proceedings of the Second University of Oklahoma Conference
77. *P. Fletcher and W. F. Lindgren,* Quasi-Uniform Spaces
78. *C.-C. Yang,* Factorization Theory of Meromorphic Functions
79. *O. Taussky,* Ternary Quadratic Forms and Norms
80. *S. P. Singh and J. H. Burry,* Nonlinear Analysis and Applications
81. *K. B. Hannsgen, T. L. Herdman, H. W. Stech, and R. L. Wheeler,* Volterra and Functional Differential Equations
82. *N. L. Johnson, M. J. Kallaher, and C. T. Long,* Finite Geometries: Proceedings of a Conference in Honor of T. G. Ostrom
83. *G. I. Zapata,* Functional Analysis, Holomorphy, and Approximation Theory
84. *S. Greco and G. Valla,* Commutative Algebra: Proceedings of the Trento Conference
85. *A. V. Fiacco,* Mathematical Programming with Data Perturbations II
86. *J.-B. Hiriart-Urruty, W. Oettli, and J. Stoer,* Optimization: Theory and Algorithms
87. *A. Figa Talamanca and M. A. Picardello,* Harmonic Analysis on Free Groups
88. *M. Harada,* Factor Categories with Applications to Direct Decomposition of Modules
89. *V. I. Istrăţescu,* Strict Convexity and Complex Strict Convexity
90. *V. Lakshmikantham,* Trends in Theory and Practice of Nonlinear Differential Equations
91. *H. L. Manocha and J. B. Srivastava,* Algebra and Its Applications
92. *D. V. Chudnovsky and G. V. Chudnovsky,* Classical and Quantum Models and Arithmetic Problems
93. *J. W. Longley,* Least Squares Computations Using Orthogonalization Methods
94. *L. P. de Alcantara,* Mathematical Logic and Formal Systems
95. *C. E. Aull,* Rings of Continuous Functions
96. *R. Chuaqui,* Analysis, Geometry, and Probability

97. *L. Fuchs and L. Salce,* Modules Over Valuation Domains
98. *P. Fischer and W. R. Smith,* Chaos, Fractals, and Dynamics
99. *W. B. Powell and C. Tsinakis,* Ordered Algebraic Structures
100. *G. M. Rassias and T. M. Rassias,* Differential Geometry, Calculus of Variations, and Their Applications
101. *R.-E. Hoffmann and K. H. Hofmann,* Continuous Lattices and Their Applications
102. *J. H. Lightbourne III and S. M. Rankin III,* Physical Mathematics and Nonlinear Partial Differential Equations
103. *C. A. Baker and L, M. Batten,* Finite Geometrics
104. *J. W. Brewer, J. W. Bunce, and F. S. Van Vleck,* Linear Systems Over Commutative Rings
105. *C. McCrory and T. Shifrin,* Geometry and Topology: Manifolds, Varieties, and Knots
106. *D. W. Kueker, E. G. K. Lopez-Escobar, and C. H. Smith,* Mathematical Logic and Theoretical Computer Science
107. *B.-L. Lin and S. Simons,* Nonlinear and Convex Analysis: Proceedings in Honor of Ky Fan
108. *S. J. Lee,* Operator Methods for Optimal Control Problems
109. *V. Lakshmikantham,* Nonlinear Analysis and Applications
110. *S. F. McCormick,* Multigrid Methods: Theory, Applications, and Supercomputing
111. *M. C. Tangora,* Computers in Algebra
112. *D. V. Chudnovsky and G. V. Chudnovsky,* Search Theory: Some Recent Developments
113. *D. V. Chudnovsky and R. D. Jenks,* Computer Algebra
114. *M. C. Tangora,* Computers in Geometry and Topology
115. *P. Nelson, V. Faber, T. A. Manteuffel, D. L. Seth, and A. B. White, Jr.,* Transport Theory, Invariant Imbedding, and Integral Equations: Proceedings in Honor of G. M. Wing's 65th Birthday
116. *P. Clément, S. Invernizzi, E. Mitidieri, and I. I. Vrabie,* Semigroup Theory and Applications
117. *J. Vinuesa,* Orthogonal Polynomials and Their Applications: Proceedings of the International Congress
118. *C. M. Dafermos, G. Ladas, and G. Papanicolaou,* Differential Equations: Proceedings of the EQUADIFF Conference
119. *E. O. Roxin,* Modern Optimal Control: A Conference in Honor of Solomon Lefschetz and Joseph P. Lasalle
120. *J. C. Díaz,* Mathematics for Large Scale Computing
121. *P. S. Milojević,* Nonlinear Functional Analysis
122. *C. Sadosky,* Analysis and Partial Differential Equations: A Collection of Papers Dedicated to Mischa Cotlar
123. *R. M. Shortt,* General Topology and Applications: Proceedings of the 1988 Northeast Conference
124. *R. Wong,* Asymptotic and Computational Analysis: Conference in Honor of Frank W. J. Olver's 65th Birthday
125. *D. V. Chudnovsky and R. D. Jenks,* Computers in Mathematics
126. *W. D. Wallis, H. Shen, W. Wei, and L. Zhu,* Combinatorial Designs and Applications
127. *S. Elaydi,* Differential Equations: Stability and Control
128. *G. Chen, E. B. Lee, W. Littman, and L. Markus,* Distributed Parameter Control Systems: New Trends and Applications
129. *W. N. Everitt,* Inequalities: Fifty Years On from Hardy, Littlewood and Pólya
130. *H. G. Kaper and M. Garbey,* Asymptotic Analysis and the Numerical Solution of Partial Differential Equations
131. *O. Arino, D. E. Axelrod, and M. Kimmel,* Mathematical Population Dynamics: Proceedings of the Second International Conference
132. *S. Coen,* Geometry and Complex Variables
133. *J. A. Goldstein, F. Kappel, and W. Schappacher,* Differential Equations with Applications in Biology, Physics, and Engineering
134. *S. J. Andima, R. Kopperman, P. R. Misra, J. Z. Reichman, and A. R. Todd,* General Topology and Applications
135. *P Clément, E. Mitidieri, B. de Pagter,* Semigroup Theory and Evolution Equations: The Second International Conference
136. *K. Jarosz,* Function Spaces
137. *J. M. Bayod, N. De Grande-De Kimpe, and J Martínez-Maurica, p*-adic Functional Analysis

*Additional Volumes in Preparation*

# *p*-adic functional analysis

Edited by
José M. Bayod
***Universidad de Cantabria***
***Santander, Spain***

N. De Grande-De Kimpe
***Vrije Universiteit Brussel***
***Brussels, Belgium***

J. Martínez-Maurica
***Universidad de Cantabria***
***Santander, Spain***

CRC Press
Taylor & Francis Group
Boca Raton London New York

CRC Press is an imprint of the
Taylor & Francis Group, an **informa** business

CRC Press
Taylor & Francis Group
6000 Broken Sound Parkway NW, Suite 300
Boca Raton, FL 33487-2742

First issued in hardback 2017



ISBN-13: 978-0-8247-8660-1 (pbk)
ISBN-13: 978-1-138-44181-1 (hbk)



**Library of Congress Cataloging-in-Publication**

*p*-adic functional analysis / edited by José M. Bayod, N. De Grande-De Kimpe, J. Martínez-Maurica.
p. cm. -- (Lecture notes in pure and applied mathematics ; 137)
Papers from an international conference held June 4-9, 1990 at Laredo, Spain and sponsored by the Universidad de Cantabria and the city of Laredo.
Includes bibliographic references and index.
ISBN 0-8247-8660-2
1. p-adic analysis--Congresses. I. Bayod, José Manuel. II. De Grande-De Kimpe, N. III. Martínez-Maurica, J. IV. Universidad de Cantabria. V. Series: Lecture notes in pure and applied mathematics ; v. 137.
QA241.P23 1991 91-41733
512'.74--dc20 CIP

**Visit the Taylor & Francis Web site at**
**http://www.taylorandfrancis.com**

**and the CRC Press Web site at**
**http://www.crcpress.com**

# PREFACE

From June 4th to 9th, 1990, an international conference on $p$-adic functional analysis was held at Laredo, Spain. In recent years there had been a growing interest in non-archimedean functional analysis which resulted in a considerable number of papers. Therefore, the organizers thought that there was a need to bring together, for the first time, some of the researchers in the field, in order to interchange ideas and make plans for future developments.

This book contains the texts of sixteen of the lectures presented at this conference, as well as a paper on open problems written by A. van Rooij and W. Schikhof. During the production of the book, two of the 19 problems discussed in detail in this paper were solved; rather than either having the paper rewritten or ignoring those answers, we decided to include them here as a sign of the vitality of the field and an encouragement for others to address the rest of the problems. These two papers are labeled as appendices, since they were not discussed at the meeting.

The conference was sponsored by the Universidad de Cantabria (Rectorado, Facultad de Ciencias and Departamento de Matemáticas, Estadística y Computación) and the City of Laredo, to whom we express our gratitude. We benefited from the constant help of Mr. Rufo de Francisco and from the efforts that Laredo is making to become a center for scientific meetings, for which it offers excellent facilities.

We also wish to thank Marcel Dekker, Inc., for producing this volume and the Mathematisch Instituut of the Katholieke Universiteit Nijmegen, The Netherlands, for their secretarial help with the manuscripts.

José M. Bayod
N. De Grande-De Kimpe
J. Martínez-Maurica

# CONTENTS

# CONTRIBUTORS

DR. J. ARAUJO
Departamento de Matemáticas
Universidad de Oviedo
Edificio Facultad de Ciencias
Calvo Sotelo s/n
33007 Oviedo, Spain

PROF. JOSÉ M. BAYOD
Facultad de Ciencias
Universidad de Cantabria
Av. de los Castros, s/n
39071 Santander, Spain

DR. SABINE BORREY
Vrije Universiteit Brussel
Faculteit Toegepaste Wetenschappen
Dienst Wiskundige Analyse
Pleinlaan 2
1050 Brussel, Belgium

PROF. N. DE GRANDE-DE KIMPE
Vrije Universiteit Brussel
Departement Wiskunde en Informatica
Pleinlaan 2 (10F7)
1050 Brussel, Belgium

DR. B. DIARRA
Université Blaise Pascal
Dep. de Mathématiques Pures
F-63177 Aubiere Cedex, France

PROF. A. ESCASSUT
Université Blaise Pascal
Dep. de Mathématiques Pures
F-63177 Aubiere Cedex, France

PROF. P. FERNÁNDEZ-FERREIRÓS
Facultad de Ciencias
Universidad de Cantabria
Av. de los Castros, s/n
39071 Santander, Spain

PROF. O. IORDACHE
Dept. Chemical Engeneering
Polytechnical Institute of Bucharest
Bucharest, CP 12-86, Romania

PROF. A.K. KATSARAS
University of Ioannina
Department of Mathematics
Ioannina 45110, Greece

PROF. J. MARTÍNEZ-MAURICA
Facultad de Ciencias
Universidad de Cantabria
Av. de los Castros, s/n
39071 Santander, Spain

PROF. S. NAVARRO
Universidad de Santiago de Chile
Departamento de Matemáticas
Facultad de Ciencia
Casilla 5659 C-2
Santiago, Chile

PROF. C. PÉREZ-GARCÍA
Facultad de Ciencias
Universidad de Cantabria
Av. de los Castros, s/n
39071 Santander, Spain

DR. G. RANGAN
University of Madras
The Ramanujan Institute
Madras - 5, India

DR. M.S. SALEEMULLAH
University of Madras
The Ramanujan Institute
Madras - 5, India

DR. M.-C. SARMANT
Université Pierre et Marie Curie
Dept. de Mathématiques
Tour 45-46, 5ème étage
4 Place Jussieu
F-75230 Paris 05, France

PROF. W.H. SCHIKHOF
Katholieke Universiteit
Mathematisch Instituut
Toernooiveld
6525 ED Nijmegen, The Netherlands

PROF. N. SHELL
The City College of New York (CUNY)
Department of Mathematics
Convent Avenue at 138th Street
New York, NY 10031, USA

PROF. L. VAN HAMME
Vrije Universiteit Brussel
Faculteit Toegepaste Wetenschappen
Dienst Wiskundige Analyse
Pleinlaan 2
1050 Brussel, Belgium

A.C.M. VAN ROOIJ
Katholieke Universiteit
Mathematisch Instituut
Toernooiveld
6525 ED Nijmegen, The Netherlands

# *p*-adic functional analysis

# DISTANCE FROM AN ISOMETRY TO THE BANACH-STONE MAPS

Jesús Araujo*

Throughout this paper $K$ denotes a complete commutative nonarchimedean valued field endowed with a nontrivial valuation. $X, Y$ are always two separated zerodimensional compact spaces. By $C(X)$ we indicate the nonarchimedean Banach space of all continuous functions from $X$ into $K$ endowed with the uniform norm.

A subset of a topological space is said to be clopen if it is both open and closed. We say that a clopen subset of $X$ is proper if it is nonempty and different from $X$. We shall denote $\mathcal{B}(X)$ the set of all the clopen subsets of $X$. If $U \in \mathcal{B}(X)$ we shall denote by $\xi_U$ the characteristic function on $U$.

**DEFINITION 0.1.** ([BN1]) A *linear isometry $T$ of $C(X)$ onto $C(Y)$ is a Banach-Stone map if there exists a homeomorphism $h$ of $Y$ onto $X$ and an $a \in C(Y)$, $|a(y)| \equiv 1$, such that $(Tf)(y) = a(y)f(h(y))$, for every $y \in Y$, $f \in C(X)$.*

Unlike the archimedean case, not every linear surjective isometry $T : C(X) \rightarrow C(X)$ is a Banach-Stone map ([AMM], Theorem 2). More generally, if $X$ is any space with more than one point, the space $BS(C(X))$ of all the Banach-Stone maps from $C(X)$ onto itself has empty interior in the space $I(C(X))$ of all the linear surjective isometries from $C(X)$ onto itself with respect to the topology defined by the norm $\|T\| := \sup\{\|Tf\| : \|f\| \leq 1\}$ ([AMM], Theorem 7).

The aim of this paper is the study of some properties of a given linear surjective isometry $T : C(X) \rightarrow C(Y)$ depending on its distance to $BS(C(X), C(Y))$, namely the space of all the Banach-Stone maps from $C(X)$ onto $C(Y)$. This will be done by characterizing in some different ways the isometries $T$ whose distance $d(T, BS(C(X), C(Y)))$ to $BS(C(X), C(Y))$ is less or equal than a given $s \in (0, 1)$. When the distance is zero, that is, when the isometry is a Banach-Stone map ([AMM],

* Research partially supported by the Spanish Dirección General de Investigación Científica y Técnica (DGICYT, PS87-0094).

Corollary 5) the characterizations we give are the obtained in [BN1] and [AMM] (Theorem 4 and Corollaries 2 and 4).

## 1. PRELIMINARY RESULTS

For $s \geq 0$ and $f \in C(X)$, we set $c_s(f) := \{x \in X : |f(x)| > s\}$. Note that for $s = 0$, $c_s(f)$ is nothing else but $c(f)$, the cozero set of $f$.

**LEMMA 1.1.** (cf. [BN1], p. 243) *Let $f \in C(X)$, $\|f\| = 1$, $s \in (0,1)$. Then there exist $m \in \mathbb{N}$, pairwise disjoint $V_1, V_2, \ldots V_m \in \mathcal{B}(X)$ and $\alpha_1, \alpha_2, \ldots, \alpha_m \in K$, verifying $|\alpha_i| \in (s,1]$ for all $i \in \{1,2,\ldots,m\}$ such that*

$$\left\| f - \sum_{i=1}^{m} \alpha_i \xi_{V_i} \right\| \leq s.$$

**LEMMA 1.2.** *Let $s \in (0,1)$. Let $T : C(X) \to C(Y)$ be a linear isometry. The following statements are equivalent:*

(1) *If $U \cap V = \varnothing$ then $c_s(T\xi_U) \cap c_s(T\xi_V) = \varnothing$ $(U, V \in \mathcal{B}(X))$.*

(2) *If $c_s(f) \cap c_s(g) = \varnothing$ then $c_s(Tf) \cap c_s(Tg) = \varnothing$ $(f, g \in C(X),\ \|f\|, \|g\| = 1)$.*

(3) *If $c(f) \cap c(g) = \varnothing$ then $c_s(Tf) \cap c_s(Tg) = \varnothing$ $(f, g \in C(X),\ \|f\|, \|g\| = 1)$.*

*Proof.* (2) $\Rightarrow$ (3) and (3) $\Rightarrow$ (1) are trivial.

Let us show (1) $\Rightarrow$ (2). Let $f, g \in C(X)$ be such that $c_s(f) \cap c_s(g) = \varnothing$, $\|f\|, \|g\| = 1$. By Lemma 1.1 there are pairwise disjoint $U_1, U_2, \ldots, U_n \in \mathcal{B}(X)$, pairwise disjoint $V_1, V_2, \ldots, V_m \in \mathcal{B}(X)$, $\alpha_1, \alpha_2, \ldots, \alpha_n \in K$ and $\beta_1, \beta_2, \ldots, \beta_m \in K$ such that

$$\left\| f - \sum_{i=1}^{n} \alpha_i \xi_{U_i} \right\| \leq s, \quad \left\| g - \sum_{j=1}^{m} \beta_j \xi_{V_j} \right\| \leq s,$$

where $|\alpha_i|, |\beta_j| \in (s,1]$ $(i \in \{1,2,\ldots,n\}, j \in \{1,2,\ldots,m\})$.

We have that $c_s(f) = c_s\left(\sum_{i=1}^{n} \alpha_i \xi_{U_i}\right) = \bigcup_{i=1}^{n} c_s(\alpha_i \xi_{U_i}) = \bigcup_{i=1}^{n} U_i$ and in the same way $c_s(g) = \bigcup_{j=1}^{m} V_j$.

Thus since $c_s(f) \cap c_s(g) = \varnothing$, then for every $i, j$, $U_i \cap V_j = \varnothing$. Therefore for every $i, j$,

$$(*) \qquad c_s(T\xi_{U_i}) \cap c_s(T\xi_{V_j}) = \varnothing.$$

On the other hand since $T$ is an isometry

$$\left\| Tf - T\left(\sum_{i=1}^{n} \alpha_i \xi_{U_i}\right) \right\| \leq s$$

and

$$\left\| Tg - T\left(\sum_{j=1}^{m} \beta_j \xi_{V_j}\right) \right\| \leq s.$$

We have that $c_s(Tf) = c_s\left(T\left(\sum_{i=1}^{n} \alpha_i \xi_{U_i}\right)\right) \subset \bigcup_{i=1}^{n} c_s(\alpha_i T\xi_{U_i}) \subset \bigcup_{i=1}^{n} c_s(T\xi_{U_i})$ and in the same way $c_s(Tg) \subset \bigcup_{j=1}^{m} c_s(T\xi_{V_j})$.
Because of $(*)$ we conclude that $c_s(Tf) \cap c_s(Tg) = \varnothing$.

**DEFINITION 1.3.** *Let $s \in (0,1)$. We shall say that a map $T : C(X) \to C(Y)$ is $s$-separating if $c(f) \cap c(g) = \varnothing$ implies that $c_s(Tf) \cap c_s(Tg) = \varnothing$, $(f, g \in C(X)$, $\|f\|, \|g\| = 1)$.*

This is a more general notion than the notion of *separating map* given in [BN2].

**LEMMA 1.4.** *Let $s \in (0,1)$. Let $T : C(X) \to C(Y)$ be an $s$-separating linear map. Then for every $U, V \in \mathcal{B}(X)$, $U \subset V$, $c_s(T\xi_U) \subset c_s(T\xi_V)$.*

*Proof.* Let $y \in Y$ be such that $|(T\xi_U)(y)| = r > s$. Since $T$ is $s$-separating,

$$\begin{aligned} |(T\xi_V)(y)| &= |(T\xi_{V-U})(y) + (T\xi_U)(y)| \\ &= |(T\xi_U)(y)| = r \end{aligned}$$

and we are done.

**LEMMA 1.5.** *Let $s \in (0,1)$. Let $T : C(X) \to C(Y)$ be an $s$-separating linear map. Let $U_1, U_2, \ldots, U_n \in \mathcal{B}(X)$. Then*

$$\bigcap_{i=1}^{n} c_s(T\xi_{U_i}) = c_s(T\xi_{U_1 \cap U_2 \cap \ldots \cap U_n}).$$

*Proof.* By Lemma 1.4 we need only prove that $\bigcap_{i=1}^{n} c_s(T\xi_{U_i}) \subset c_s(T\xi_{U_1 \cap U_2 \cap \ldots \cap U_n})$. For that we shall proceed by induction over $n$.

First we consider the case $n = 2$. We have that $U_1$ can be expressed as the disjoint union $U_1 = (U_1 - U_2) \cup (U_1 \cap U_2)$. Let $y \in c_s(T\xi_{U_1}) \cap c_s(T\xi_{U_2})$, this is, $|(T\xi_{U_1})(y)| > s$, $|(T\xi_{U_2})(y)| > s$. Since $U_2 \cap (U_1 - U_2) = \varnothing$, $|(T\xi_{U_1 - U_2})(y)| \leq s$, and then

$$s < |(T\xi_{U_1})(y)| = |(T\xi_{U_1 - U_2})(y) + (T\xi_{U_1 \cap U_2})(y)|$$

from which it can be deduced that $y \in c_s(T\xi_{U_1 \cap U_2})$.

Now suppose that the result holds for $n = k$, that is,

$$\bigcap_{i=1}^{k} c_s(T\xi_{U_i}) \subset c_s(T\xi_{U_1 \cap U_2 \cap \ldots \cap U_k}).$$

Let us prove the result for $n = k+1$. It now follows that

$$\begin{aligned} \bigcap_{i=1}^{k+1} c_s(T\xi_{U_i}) &= \left( \bigcap_{i=1}^{k} c_s(T\xi_{U_i}) \right) \cap c_s(T\xi_{U_{k+1}}) \\ &\subset c_s(T\xi_{U_1 \cap U_2 \cap \ldots \cap U_k}) \cap c_s(T\xi_{U_{k+1}}) \\ &\subset c_s(T\xi_{U_1 \cap U_2 \cap \ldots \cap U_k \cap U_{k+1}}). \end{aligned}$$

**COROLLARY 1.6.** *Let $s \in (0,1)$. Let $T : C(X) \to C(Y)$ be an $s$-separating linear map. If $U_1, U_2, \ldots, U_n \in \mathcal{B}(X)$ are such that $U_1 \cap U_2 \cap \ldots \cap U_n = \varnothing$, then $\bigcap_{i=1}^{n} c_s(T\xi_{U_i}) = \varnothing$.*

**LEMMA 1.7.** *Let $s \in [0,1)$. Let $T : C(X) \to C(Y)$ be a linear surjective isometry such that $d(T, BS(C(X), C(Y))) > s$. Then for each homeomorphism $h : Y \to X$ there exists a proper $U \in \mathcal{B}(X)$ such that*

$$\|T\xi_U - (T\xi_X)(\xi_U \circ h)\| > s.$$

*Proof.* First I shall prove that there exists an $f \in C(X)$ such that $\|f\| = 1$ and $\|Tf - (T\xi_X)(f \circ h)\| > s$. This will be done by considering two different possibilities.
(a) If there exists $y_0 \in Y$ such that $|(T\xi_X)(y_0)| < 1$, then we set $f := T^{-1}\xi_Y$.
For this choice of $f$ we have

$$\begin{aligned} |(Tf)(y_0) - (T\xi_X)(y_0) f(h(y_0))| &= |\xi_Y(y_0) - (T\xi_X)(y_0) f(h(y_0))| \\ &= |1 - (T\xi_X)(y_0) f(h(y_0))| = 1 \end{aligned}$$

because $|(T\xi_X)(y_0)| < 1$ and $|f(h(y_0))| \leq 1$.

So $\|Tf - (T\xi_X)(f \circ h)\| > s$.
(b) If $|(T\xi_X)(y)| = 1$ for every $y \in Y$, then $S : C(X) \to C(Y)$ defined by $(Sf)(y) := (T\xi_X)(y)f(h(y))$ for every $f \in C(X)$ and $y \in Y$, is a Banach-Stone map and then $\|T-S\| > s$ which implies that there exists $f \in C(X)$, $\|f\|=1$ and $\|(Tf-(T\xi_X)(f \circ h)\| > s$.

Then in any case there exists $f \in C(X)$ such that $\|f\|=1$ and $\|Tf-(T\xi_X)(f \circ h)\| > s$. Now let $\{\xi_{U_i} : i \in I\}$ be an orthonormal base of $C(X)$ with $U_i \in \mathcal{B}(X)$ for all $i \in I$. Then $f = \sum_{i \in I} a_i \xi_{U_i}$ where $a_i \in K$, $|a_i| \leq 1$ for all $i \in I$.

So we have

$$\left\| T\left(\sum_{i \in I} a_i \xi_{U_i}\right) - (T\xi_X)\left(\sum_{i \in I} a_i \xi_{U_i} \circ h\right)\right\| = \left\| \sum_{i \in I} a_i (T\xi_{U_i} - (T\xi_X)(\xi_{U_i} \circ h))\right\| > s$$

and then there exists $i \in I$ such that

$$|a_i| \, \|T\xi_{U_i} - (T\xi_X)(\xi_{U_i} \circ h)\| > s.$$

Then $\|T\xi_{U_i} - (T\xi_X)(\xi_{U_i} \circ h)\| > s$. Also it is clear that this $U_i$ has to be proper.

**LEMMA 1.8.** *Let $s \in (0,1)$. Let $T : C(X) \to C(Y)$ be an $s$-separating linear surjective isometry. Then $c_s(T\xi_X) = Y$.*

*Proof.* If $c_s(T\xi_X) \neq Y$, set $f := T^{-1}\xi_{Y - c_s(T\xi_X)}$. Because of Lemma 1.1, there exist $V_1, V_2, \ldots V_m \in \mathcal{B}(X)$ and $\alpha_1, \alpha_2, \ldots, \alpha_m \in K$, $|\alpha_i| \in (s,1]$ for all $i \in \{1,2,\ldots,m\}$ such that

$$\left\| f - \sum_{i=1}^{m} \alpha_i \xi_{V_i} \right\| \leq s.$$

Then $\left\| \xi_{Y - c_s(T\xi_X)} - \sum_{i=1}^{m} \alpha_i T\xi_{V_i} \right\| \leq s$. If $y \notin c_s(T\xi_X)$, $\left| \sum_{i=1}^{m} \alpha_i (T\xi_{V_i})(y) \right| = 1$ and hence there exists $j \in \{1,2,\ldots,m\}$ such that $|(T\xi_{V_j})(y)| = 1$. Then by Lemma 1.4 $y \in c_s(T\xi_X)$ which is a contradiction.

**LEMMA 1.9.** *Suppose that the valuation of $K$ is dense or that the residue class field of $K$ is different from $\{0,1\}$ and let $s \in (0,1)$. Let $a, b, c \in K$ be such that $s < |a| \leq |b| \leq 1$, $|c| \leq 1$. Then there are $\alpha, \beta \in K$, $|\alpha|, |\beta| \in (s,1]$ such that $\alpha a + \beta b + ca = 0$.*

*Proof.* Let us choose $\alpha \in K$ such that $|\alpha| \in (s,1]$ and $|\alpha + c| = 1$. Then $\alpha$ and $\beta := -(\alpha + c)ab^{-1}$ have the required properties.

**LEMMA 1.10.** *Let $\alpha, \beta, \gamma \in K$ be such that $|\alpha|, |\beta| = 1$, $|\gamma| < 1$. Then there exists $\delta \in K$, $|\delta| = 1$ with $\gamma = \alpha + \delta\beta$.*

*Proof.* It is easy by taking $\delta := (\gamma - \alpha)\beta^{-1}$.

## 2. THE MAIN RESULTS

The means to prove the main results 2.2 and 2.3 are developed in 2.1.

**THEOREM 2.1.** *Let $s \in (0,1)$. Let $T : C(X) \to C(Y)$ be a linear surjective isometry. If $T$ is $s$-separating then there exists a surjective homeomorphism $h : Y \to X$ such that $h(c_s(T\xi_U)) = U$ for every $U \in \mathcal{B}(X)$.*

*Proof.* Let $y \in Y$. By 1.8, $y \in c_s(T\xi_X)$ so we can define

$$\mathcal{A}(y) := \bigcap_{\substack{U \in \mathcal{B}(X) \\ y \in c_s(T\xi_U)}} U.$$

Notice that $\mathcal{A}(y) \subset U$ if $U \in \mathcal{B}(X)$ is such that $y \in c_s(T\xi_U)$.

Let us suppose that $y \in c_s(T\xi_{U_1}) \cap c_s(T\xi_{U_2}) \cap \ldots \cap c_s(T\xi_{U_n})$. Then by Corollary 1.6 $U_1 \cap U_2 \cap \ldots \cap U_n \neq \emptyset$. Since $X$ is compact we have that $\mathcal{A}(y) \neq \emptyset$. On the other hand let us see that $\mathcal{A}(y)$ has only one point $x$.

Let $x \in \mathcal{A}(y)$ and let $x' \neq x$. Let $V \in \mathcal{B}(X)$ be such that $x \in V$, $x' \notin V$ and $U \in \mathcal{B}(X)$ be such that $y \in c_s(T\xi_U)$. Thus

$$|(T\xi_U)(y)| = |(T\xi_{U\cap V})(y) + (T\xi_{U\cap(X-V)})(y)| > s.$$

We know that $\mathcal{A}(y)$ is not contained in $U \cap (X-V)$ so that $y \notin c_s(T\xi_{U\cap(X-V)})$. Hence $|(T\xi_{U\cap V})(y)| > s$ and $y \in c_s(T\xi_{U\cap V})$. In particular $\mathcal{A}(y) \subset U \cap V$ and $x' \notin \mathcal{A}(y)$.

For each $y \in Y$ we shall denote as $h(y)$ the only element of $\mathcal{A}(y)$. It is obvious that if $U \in \mathcal{B}(X)$, then $h(c_s(T\xi_U)) \subset U$.

Now we shall prove that the map $h$ is a homeomorphism of $Y$ onto $X$.

1) *$h$ is injective.* Let $y, y' \in Y$, $y \neq y'$. Let us see that $h(y) \neq h(y')$. Let $U(y)$ be a clopen neighbourhood of $y$ such that $y' \notin U(y)$. Set $f := T^{-1}\xi_{U(y)}$. Because of Lemma 1.1, $f$ can be put as

$$f = \sum_{i=1}^{n} \alpha_i \xi_{V_i} + g,$$

where $\|g\| \leq s$, $V_i \cap V_j = \emptyset$ if $i \neq j$, $i,j \in \{1,2,,\ldots,n\}$ and $|\alpha_i| \in (s,1]$ for every $i \in \{1,2,\ldots,n\}$. Since $1 = |(Tf)(y)| = |\sum_{i=1}^{n} \alpha_i(T\xi_{V_i}(y)|$, then there exists $j \in \{1,2,\ldots,n\}$ such that $|\alpha_j| = 1$ and $y \in c_s(T\xi_{V_j})$. It is easy to check that $y' \notin c_s(T\xi_{V_j})$ and, since by Lemma 1.8 $y' \in c_s(T\xi_X)$, then $y' \in c_s(T\xi_{X-V_j})$. So $h(y) \in V_j$ and $h(y') \notin V_j$.

2) *$h$ is surjective.* For a given $x \in X$, set

$$\mathcal{A}'(x) := \bigcap_{\substack{x \in U \\ U \in \mathcal{B}(X)}} c_s(T\xi_U).$$

Let us see that for each $x \in X$, $\mathcal{A}'(x) \neq \emptyset$. In order to prove that, let us consider $U_1, U_2, \ldots, U_n \in \mathcal{B}(X)$ such that $x \in \bigcap_{i=1}^{n} U_i$. By Lemma 1.5,

$$\bigcap_{i=1}^{n} c_s(T\xi_{U_i}) = c_s(T\xi_{U_1 \cap U_2 \cap \ldots \cap U_n}) \neq \emptyset.$$

Since $Y$ is compact and $c_s(T\xi_U)$ is closed for any $U \in \mathcal{B}(X)$, then $\mathcal{A}'(x) \neq \emptyset$.
Let us see that if $y \in \mathcal{A}'(x)$, then $h(y) = x$. Let $U(x)$ be a clopen neighborhood of $x$. Since $y \in \mathcal{A}'(x)$, $y \in c_s(T\xi_{U(x)})$ and then $h(y) \in U(x)$. So $h(y) \in \bigcap_{\substack{U \in \mathcal{B}(X) \\ x \in U}} U$, this is, $h(y) = x$.

3) *$h$ is continuous.*

Let $y \in Y$ and $x = h(y)$. Let $U(x)$ be a clopen neighbourhood of $x$. Let us see that $y \in c_s(T\xi_{U(x)})$. Otherwise, since $c_s(T\xi_X) = Y$ by Lemma 1.8, then $y \in c_s(T\xi_{X-U(x)})$; so $h(y) = x \notin U(x)$, which is a contradiction. Thus since $h(c_s(T\xi_{U(x)})) \subset U(x)$, $h$ is continuous and consequently a homeomorphism.

Finally let us see that $h(c_s(T\xi_U)) = U$ for every $U \in \mathcal{B}(X)$. Let $x \in U, U \in \mathcal{B}(X)$. Since $h^{-1}(x) \in \mathcal{A}'(x) \subset c_s(T\xi_U)$, then $U \subset h(c_s(T\xi_U))$. The other inclusion has already been seen.

**THEOREM 2.2.** *Let $s \in (0,1)$. Let $T : C(X) \to C(Y)$ be a linear surjective isometry. Then the following statements are equivalent,*
(1) *$d(T, BS(C(X), C(Y))) \leq s$.*
(2) *$T$ is $s$-separating.*
*Moreover if the valuation of $K$ is dense or its residue class field is different from $\{0,1\}$, then (1) and (2) are also equivalent to*

(3) *If* $f \in C(X)$ *and* $1 \geq |f(x)| > s$ *for every* $x \in X$*, then* $Tf$ *is invertible.*

*Proof.* (1) $\Rightarrow$ (2). Let us suppose that $U, V \in \mathcal{B}(X)$ are disjoint. If $y \in c_s(T\xi_U) \cap c_s(T\xi_V)$, then $|(T\xi_U)(y)|, |(T\xi_V)(y)| \geq r > s$. By hypothesis there exists a Banach-Stone map $S$ such that $\|S-T\| < r$. It is easy to check that $y \in c_s(S\xi_U) \cap c_s(S\xi_V)$. However this is impossible because $S$ is a Banach-Stone map ([BN1], p.243).

(2) $\Rightarrow$ (1). By Theorem 2.1 we know that there exists a homeomorphism $h : Y \to X$ satisfying $h(c_s(T\xi_U)) = U$ for every $U \in \mathcal{B}(X)$. If $d(T, BS(C(X), C(Y))) > s$, then by Lemma 1.7 there exists a proper $U \in \mathcal{B}(X)$ such that

$$\|T\xi_U - (T\xi_X)(\xi_U \circ h)\| = \|T\xi_U - (T\xi_X)\xi_{h^{-1}(U)}\| > s.$$

If $y \notin h^{-1}(U) = c_s(T\xi_U)$ we have that

$$|(T\xi_U)(y) - (T\xi_X)(y)\xi_{h^{-1}(U)}(y)| = |(T\xi_U)(y)| \leq s,$$

and thus there exists $y_0 \in h^{-1}(U)$ satisfying

$$|(T\xi_U)(y_0) - (T\xi_X)(y_0)\xi_{h^{-1}(U)}(y_0)| = |(T\xi_U)(y_0) - (T\xi_X)(y_0)| > s.$$

So

$$|(T\xi_{X-U})(y_0)| = |(T\xi_X)(y_0) - (T\xi_U)(y_0)| > s$$

and therefore $y_0 \in c_s(T\xi_{X-U})$ which contradicts that $T$ is $s$-separating.

(1) $\Rightarrow$ (3). Let $f \in C(X)$, $1 \geq |f(x)| > s$ for every $x \in X$. Since $X$ is compact and nonempty we have that $r := \inf\{|f(x)| : x \in X\} > s$. Let $S \in BS(C(X), C(Y))$ such that $\|T-S\| < r$.

Since $S$ is a Banach-Stone map

$$\{|(Sf)(y)| : y \in Y\} = \{|f(x)| : x \in X\} \subset [r, 1]$$

by ([AMM], Theorem 4 and Corollary 2) and hence for any $y \in Y$, $|(Sf)(y)| \geq r$.

Moreover, $|(Tf)(y) - (Sf)(y)| < r$ and hence $|(Tf)(y)| = |(Sf)(y)| \geq r$ for every $y \in Y$.

Thus $Tf$ is invertible.

(3) $\Rightarrow$ (2). Suppose that (2) does not hold. Then, by Lemma 1.2, there are disjoint subsets $U, V \in \mathcal{B}(X)$ such that

$$c_s(T\xi_U) \cap c_s(T\xi_V) \neq \varnothing.$$

Let $y \in c_s(T\xi_U) \cap c_s(T\xi_V)$. Without loss of generality we can suppose that $|(T\xi_U)(y)| \leq |(T\xi_V)(y)|$.

By Lemma 1.9 there exist elements $\alpha, \beta \in K$, $1 \geq |\alpha|, |\beta| > s$, such that

$$\alpha(T\xi_U)(y) + \beta(T\xi_V)(y) + (T\xi_{X-(U\cup V)})(y)(T\xi_U)(y) = 0.$$

Set

$$f := \alpha\xi_U + \beta\xi_V + (T\xi_U)(y)\xi_{X-(U\cup V)} \in C(X).$$

The $\|f\| \leq 1$. Also it is easy to check that $1 \geq |f(x)| > s$ for every $x \in X$.

Moreover $(Tf)(y) = 0$ and thus $Tf$ is not invertible.

In the previous theorem, we have included some additional conditions in order to prove the equivalence between (1) and (3). Obviously (1) implies (3) without assuming any extra condition on $K$, but the converse generally does not hold, as the following theorem shows.

**THEOREM 2.3.** *Assume that $K$ is endowed with a discrete valuation and that its residue class field is $\{0, 1\}$. If $X$ has more than two points, then there are $s \in (0, 1)$ and a linear surjective isometry $T : C(X) \to C(X)$ satisfying*

(1) *$d(T, BS(C(X))) = 1$.*

(2) *If $f \in C(X)$ and $1 \geq |f(x)| > s$ for every $x \in X$, then $Tf$ is invertible.*

*Proof.* Let $a, b \in X$, $a \neq b$. Let $U$ be a proper clopen subset of $X$ such that $a, b \notin U$. Let us define $T : C(X) \to C(X)$ as

$$Tf := f + (f(a) + f(b))\xi_U$$

for every $f \in C(X)$. $T$ is a linear continuous map from $C(X)$ into itself and $\|T\| \leq 1$. One easily infers that $T$ has an inverse, viz.

$$f \to f - (f(a) + f(b))\xi_U.$$

So $T$ is a linear surjective isometry.

On the other hand, let $s \in (\pi, 1)$, where $\pi := \sup\{|\alpha| : \alpha \in K, |\alpha| < 1\}$ and let $f \in C(X)$. If, for every $x \in X$, $|f(x)| \in (s, 1]$, then $|f(x)| = 1$ for every $x \in X$. Since the residue class field is $\{0, 1\}$, then for every $x \in X$, $|(Tf)(x)| = 1$, and so $Tf$ is invertible.

Moreover we shall see that $d(T, BS(C(X))) = 1$. Let $U(a), U(b)$ be disjoint clopen neighbourhoods of $a$ and $b$ respectively such that $U \cap U(a) = \emptyset = U \cap U(b)$. We have that

$$|(T\xi_{U(a)})(y)| = 1 = |(T\xi_{U(b)})(y)|$$

for each $y \in U$ and then, by Theorem 2.2, $d(T, BS(C(X))) = 1$.

## 3. SOME OTHER CHARACTERIZATIONS

If $s \geq 0$ and $f \in C(X)$, we set $v_s(f) := \{|f(x)| : x \in X\} - [0, s]$. Notice that $\nu_0(f)$ coincides with $v(f) = \{|f(x)| : x \in X\} - \{0\}$ as defined in [AMM].

Also we shall denote by $\mathbb{J}$ the interval $(0, \pi)$ where $\pi = \sup\{|x| : x \in K, |x| < 1\}$.

**THEOREM 3.1.** *Let $s \in \mathbb{J}$. If $T : C(X) \to C(Y)$ is a linear surjective isometry, then the following statements are equivalent:*

(1) *$d(T, BS(C(X), C(Y))) \leq s$.*

(2) *$v_s(Tf) \subset v_s(f)$ for every $f \in C(X)$ such that $\|f\| \leq 1$.*

(3) *$\inf v_s(f) \leq \inf v_s(Tf)$ for every $f \in C(X)$ such that $\|f\| \leq 1$.*

(4) *If $v_s(f) = \{1\}$, then $v_s(Tf) = \{1\}$, $(f \in C(X))$.*

*Proof.* (1) $\Rightarrow$ (2). Let $f \in C(X)$, $\|f\| \leq 1$. Let $r \in v_s(Tf)$. Then there is $y \in Y$ such that $|(Tf)(y)| = r$. Since $s < r$, $d(T, BS(C(X), C(Y))) < r$ and we can choose a Banach-Stone map $S : C(X) \to C(Y)$ such that $\|S-T\| < r$ . Then we shall have

$$\begin{aligned} |(Sf)(y)| &= |((S-T)(f))(y) + (Tf)(y)| \\ &= |(Tf)(y)| = r. \end{aligned}$$

So $r \in v_s(Sf)$. Since $S$ is a Banach-Stone map, $v(Sf) = v(f)$ ([AMM], Corollary 2) and hence $r \in v(f)$ and (2) is proved.

(2) $\Rightarrow$ (3) is obvious.

(3) $\Rightarrow$ (4). Let us suppose that $v_s(f) = \{1\}$. Then it is obvious that $\|f\| = 1$ and we have

$$\sup v_s(Tf) = \sup v_s(f) = 1 = \inf v_s(f) \leq \inf v_s(Tf).$$

Thus $v_s(Tf) = \{1\}$ and (4) is satisfied.

(4) $\Rightarrow$ (1). We prove the contrapositive. Let us suppose that (1) is not satisfied. Then, by Theorem 2.2, $T$ is not $s$-separating. By Lemma 1.2 there are two

disjoint proper clopen subsets $U$, $V$ and an element $y \in Y$ such that $|(T\xi_U)(y)| > s$, $|(T\xi_V)(y)| > s$. By hypothesis $|(T\xi_U)(y)| = 1 = |(T\xi_V)(y)|$ and then by Lemma 1.10 there exists $\alpha \in K$, $|\alpha| = 1$, such that $s < |(T\xi_U + \alpha T\xi_V)(y)| < 1$.

Set $f := \xi_U + \alpha\xi_V$. Then $v_s(f) = \{1\}$ but

$$|(T\xi_U + \alpha T\xi_V)(y)| \in v_s(f),$$

which is the negation of (4).

**LEMMA 3.2.** *Let $T : C(X) \to C(Y)$ be a linear surjective isometry. Then $d(T, BS(C(X), C(Y))) = d(T^{-1}, BS(C(Y), C(X)))$.*

*Proof.* For every $S \in BS(C(X), C(Y))$ we have that $S^{-1} \in BS(C(Y), C(X))$ ([AMM], Remark to Theorem 4) and also

$$\begin{aligned} \|S^{-1} - T^{-1}\| &= \|T^{-1}(T - S)S^{-1}\| \\ &\leq \|T - S\|. \end{aligned}$$

Thus $d(T^{-1}, BS(C(Y), C(X))) \leq d(T, BS(C(X), C(Y)))$. In the same way we can obtain the other inequality.

**COROLLARY 3.3.** *Let $s \in \mathbb{J}$. If $T : C(X) \to C(Y)$ is a linear surjective isometry, the following statements are equivalent:*

(1) $d(T, BS(C(X), C(Y))) \leq s$.

(2) $\inf v_s(Tf) \leq \inf v_s(f)$ *for every* $f \in C(X)$ *such that* $\|f\| \leq 1$.

(3) $\inf v_s(f) = \inf v_s(Tf)$ *for every* $f \in C(X)$ *such that* $\|f\| \leq 1$.

(4) *If* $v_s(Tf) = \{1\}$ *then* $v_s(f) = \{1\}$, $(f \in C(X))$.

(5) $v_s(f) \subset v_s(Tf)$ *for every* $f \in C(X)$ *such that* $\|f\| \leq 1$.

(6) $v_s(Tf) = v_s(f)$ *for every* $f \in C(X)$ *such that* $\|f\| \leq 1$.

*Proof.* It follows easily from Theorem 3.1 and Lemma 3.2.

**THEOREM 3.4.** *Let $s \in \mathbb{J}$. If $T : C(X) \to C(Y)$ is a linear surjective isometry then the following statements are equivalent:*

(1) $d(T, BS(C(X), C(Y))) \leq s$.

(2) *If* $v_s(f) \subset v_s(g)$, *then* $v_s(Tf) \subset v_s(Tg)$ $(f, g \in C(X),\ \|f\|,\ \|g\| \leq 1)$.

(3) *If* $v_s(f) = v_s(g)$, *then* $v_s(Tf) = v_s(Tg)$ $(f, g \in C(X),\ \|f\|,\ \|g\| \leq 1)$.

(4) *If* $\inf v_s(f) \leq \inf v_s(g)$, *then* $\inf v_s(Tf) \leq \inf v_s(Tg)$ $(f, g \in C(X),\ \|f\|,\ \|g\| \leq 1)$.

(5) *If* $\inf v_s(f) = \inf v_s(g)$, *then* $\inf v_s(Tf) = \inf v_s(Tg)$ $(f, g \in C(X), \|f\|, \|g\| \leq 1)$.

*Proof.* (2) $\Rightarrow$ (3) and (4) $\Rightarrow$ (5) are obvious.

(1) $\Rightarrow$ (2). Suppose that $f, g \in C(X)$, $\|f\|$, $\|g\| \leq 1$ are such that $v_s(f) \subset v_s(g)$. Thus by Corollary 3.3 $v_s(Tf) = v_s(f) \subset v_s(g) = v_s(Tg)$.

(1) $\Rightarrow$ (4). Suppose that $\inf v_s(f) \leq \inf v_s(g)$. Thus by Corollary 3.3 $\inf v_s(Tf) = \inf v_s(f) \leq \inf v_s(g) = \inf v_s(Tg)$.

Let us finally see that (3) $\Rightarrow$ (1) and (5) $\Rightarrow$ (1).

Set $g := T^{-1}\xi_Y \in C(X)$ and set $g' := g\xi_A$, where $A = \{x \in X : |g(x)| = 1\}$. Obviously for every $y \in Y$,

$$|(Tg')(y)| = |(Tg)(y) - (T(g\xi_{X-A}))(y)| = 1$$

because $\|g\xi_{X-A}\| < 1$.

Thus $v_s(g') = \{1\}$ and $v_s(Tg') = \{1\}$. Then if (3) is satisfied and $v_s(f) = \{1\} = v_s(g')$, it must be that $v_s(Tf) = \{1\} = v_s(Tg')$ and then $d(T, BS(C(X), C(Y))) \leq s$ by Theorem 3.1. (5) $\Rightarrow$ (1) can be proved in the same way.

I am grateful to Prof. J. Martínez-Maurica for his valuable suggestions.

## REFERENCES

[AMM] J. Araujo and J. Martínez-Maurica. *The nonarchimedean Banach-Stone theorem.* In *p-adic Analysis, Proceedings, Trento 1989*, edited by F. Baldassarri, S. Bosch and B. Dwork, Springer-Verlag (Lect. Notes 1454), Berlin-Heidelberg-New York (1990), 64–79.

[BN1] E. Beckenstein and L. Narici. *A nonarchimedean Stone-Banach theorem.* Proc. Amer. Math. Soc. **100** (1987), 242–246.

[BN2] E. Beckenstein and L. Narici. *Automatic continuity of certain linear isomorphisms.* Acad. Royale Belg. Bull. Cl. Sci. (5) **73** (1987), 191–200.

[vR] A.C.M. van Rooij. *Non-archimedean functional analysis.* Marcel Dekker, New York, 1978.

DEPARTAMENTO DE MATEMÁTICAS, UNIVERSIDAD DE OVIEDO,
EDIFICIO FACULTAD DE CIENCIAS,
CALVO SOTELO S/N 33007, OVIEDO, SPAIN

# PSEUDOCOMPACT AND P-SPACES IN NON-ARCHIMEDEAN FUNCTIONAL ANALYSIS

Jesús Araujo, P. Fernández-Ferreirós and J. Martínez-Maurica*

This paper deals with two classes of zerodimensional topological spaces: pseudocompact and $P$-spaces. These are highly interesting in the study of the non-archimedean spaces of continuous functions. The reasons for studying these two properties together is that they can be considered to some extent (see 1.7 and 2.8) as dual properties.

Throughout $X$ is a separated topological space and $K$ a complete non-archimedean valued field endowed with a non-trivial valuation.

By $X^{\delta}$ we denote the $\mathbb{N}$-compactification and by $X^{\xi}$ the Banaschewski compactification of a zerodimensional space $X$. These can be considered to be the zerodimensional counterparts of the realcompactification $\nu X$ and the Stone-Cech compactification $\beta X$ respectively. For details concerning these compactifications we refer to [Vr].

## 1. ZERODIMENSIONAL PSEUDOCOMPACT SPACES

Recall that $X$ is said to be *pseudocompact* if every real continuous map on $X$ is bounded.

Since in the non-archimedean context we are dealing with maps whose range is included in $K$ it is natural to introduce what G. Bachman, E. Beckenstein, L. Narici and S. Warner call *K-pseudocompact* spaces defined in the following way: $X$ is said to be $K$-pseudocompact if every continuous map $f : X \to K$ is bounded.

It is clear that every pseudocompact space is $K$-pseudocompact (compose with the valuation map) but the converse is not true in general. Take for instance $X = \mathbb{R}$;

* Research partially supported by DGICYT (grant PS87-0094)

$X$ is not pseudocompact but it is $K$-pseudocompact because for every continuous $K$-valued map $f$, $f(X)$ is connected and hence a singleton; however for zerodimensional spaces we have:

**THEOREM 1.1.** *Let $X$ be zerodimensional and let $E$ be a metric space endowed with an unbounded distance. Then the following are equivalent:*

(a) *$X$ is pseudocompact.*
(b) *There is no infinite open partition of $X$.*
(c) *Every continuous map $f : X \to \mathbb{N}$ is bounded.*
(d) *Every continuous map $f : X \to E$ is bounded.*
(e) *$X$ is ultra countably compact (i.e. every countable clopen covering of $X$ has a finite subcovering).*

### Remarks and Examples

**1.2.** This result is essentially due to R.S. Pierce who proved (a) $\Leftrightarrow$ (b) $\Leftrightarrow$ (e) ([P] lemma 1.8.2 and lemma 1.9.3). Also, in the review of [Ba] in the Mathematical Reviews, P. Nyikos ([N]) observes that (d) is also a equivalent property. The other implications are easy to prove.

**1.3.** The Tychonoff plank $[0, \omega] \times [0, \Omega] - \{(\omega, \Omega)\}$ is a zerodimensional pseudocompact space which is not countably compact ([D], p.231).

**1.4.** Unlike the case of compact spaces, not every zerodimensional pseudocompact space is strongly zerodimensional. This is the case of the example which appears in [G], p.129. This zerodimensional pseudocompact space $X$ contains an infinite discrete subset $Y$ so that there are continuous unbounded real maps defined on $Y$. These maps cannot be extended continuously to the whole space $X$. In particular $X$ cannot be strongly zerodimensional ([E], theorem 3.1).

Because of the above theorem $X$ IS HEREAFTER ASSUMED TO BE A ZERODIMENSIONAL SPACE.

Recall that $X$ is non-archimedean if it has a base $\mathcal{B}$ such that if $A, B \in \mathcal{B}$, then either $A \subset B$ or $B \supset A$ or $A \cap B = \emptyset$. Every non-archimedean space is zerodimensional.

**THEOREM 1.5.** *Let $X$ be a non-archimedean pseudocompace space. Then $X$ is compact (and hence ultrametrizable).*

*Proof.* Indeed every non-archimedean space is paracompact. Also every paracompact pseudocompact space is compact ([E], theorem 3.10.22 and 5.1.11). Now the theorem follows because every non-archimedean compact space is ultrametrizable ([N2], corollary 9).

The following corollary is the zerodimensional counterpart to the well known result which says that $\nu X = \beta X$ for any pseudocompact completely regular space $X$.

**COROLLARY 1.6.** *$X$ is pseudocompact if and only if $X^\delta = X^\xi$.*

*Proof.* One implication (from the left to the right) is easy (see also [Vr], p.44). As for the other, assume $X^\delta = X^\xi$ and let $f : X \to \mathbb{N}$ be a continuous map. By [Vr], (p.42) $f$ can be extended to a continuous map $f^\delta : X^\delta \to \mathbb{N}$. Since $X^\delta$ is compact, so is $f^\delta(X^\delta)$ and hence $f$ is bounded. By 1.1 ((c) $\Rightarrow$ (a)), $X$ is pseudocompact.

This corollary had also been proved by G. Bachman, E. Beckenstein, L. Narici and S. Warner in [Ba] under the additional hypothesis of being $K$ realcompact, this is when $K$ is $\mathbb{N}$-compact. It is also well known that this last property holds if and only if the residual field $k$ is non measurable.

## APPLICATIONS

### (I) BASES IN $BC_S(X)$

As usual $BC(X)$ denotes the Banach space of all bounded continuous maps from $X$ into $K$ endowed with the supremum norm.

Following [Vr], we denote by $BC_S(X)$ the closed subspace of $BC(X)$ consisting of these elements $f$ such that $f(X)$ has a non measurable cardinal.

**COROLLARY 1.7.** *If the valuation on $K$ is dense, then $BC_S(X)$ has an (orthogonal) basis if and only if $X$ is pseudocompact.*

*Proof.* According to the theorem 2.20 (iii) in [Vr], $BC_S(X)$ is linearly isometric to $BC_S(X^\delta)$ and this last space coincides with $BC(X^\delta)$ because $X^\delta$ is $\mathbb{N}$-compact (see also p.243 in [Vr]). This space has an (orthogonal) basis if and only if $X^\delta$ is pseudocompact ([Vr], p.189). By 1.6 this last property holds if and only if $X$ is pseudocompact.

(II) LOCAL COMPACTOIDS IN $PC(X)$

As usual $PC(X)$ stands for the subspace of $BC(X)$ consisting of those maps whose range is precompact.

In a previous paper ([M]) with S. Navarro we proved the following characterization of compactoids in $PC(X)$: *A subset* $\mathcal{F} \subset PC(X)$ *is compactoid if and only if* (A) $\mathcal{F}(x)$ *is bounded in* $K$ *for every* $x \in X$ *and* (B) *For every* $\varepsilon > 0$ *there exists a finite clopen partition* $X_1, \ldots, X_n$ *of* $X$ *such that* $x, y \in X_i \Rightarrow |f(x) - f(y)| \leq \varepsilon$ $(i = 1, \ldots, n)$. Property (B) implies equicontinuity of $\mathcal{F}$; also if $X$ is pseudocompact (B) is the same as equicontinuity.

It seems reasonable (by dropping the hypothesis of boundedness in both sides) to investigate the relationship between the properties of being *locally compactoid* and *equicontinuous.*

Recall that a subset $A$ of a normed space is locally compactoid if every bounded subset of $A$ is compactoid.

**1.8.** FACT A. Let $X$ be any infinite compact subset of $K$. Define $I : X \to K$ by $i(x) = x$ $(x \in X)$. Then, $i \in PC(X)$. Also $A = [i]$ (the linear hull of $i$) is a locally compactoid subset of $PC(X)$ which fails to be equicontinuous in the accumulation points of $X$.

**1.9.** FACT B. If $X$ is pseudocompact, then every equicontinuous subset of $PC(X)$ is locally compactoid.

*Proof.* Let $A$ be an equicontinuous subset of $PC(X)$ and let $B$ be a bounded subset of $A$. Define in $X$ the equivalence relation $R$ by

$$xRy := |f(x) - f(y)| < \varepsilon, \quad \forall f \in B.$$

The equivalence classes for $R$ are clopen subsets of $X$ (here we use the equicontinuity of $A$). Then the set of all these classes is a clopen partition of $X$. Since $X$ is pseudocompact, there is only a finite number of these classes. By the above characterization of compactoids in $PC(X)$ it follows that $B$ is compactoid.

**1.10.** FACT C. If every equicontinuous subset of $PC(X)$ is locally compactoid then $X$ is pseudocompact.

*Proof.* Assume $X$ is not pseudocompact. Then, there exists an infinite partition of nonempty clopen sets $\{A_n : n \in \mathbb{N}\}$. If $k_n$ stands for the characteristic function on $A_n$, it is clear that $A = \overline{\text{co}\{k_n : n \in \mathbb{N}\}}$ is equicontinuous. However $A$ is not locally compactoid because $(k_n)$ is an orthogonal sequence in $A$ that does not tend to zero ([Vr], theorem 6.7).

## 2. *P*-SPACES

We say that a map $f : X \to K$ is locally constant if each element of $X$ possesses a neighborhood in which $f$ is constant.

It is well known ([Sch], lemma 1.2) that when $X$ is a zerodimensional space, then each continous map $f : X \to K$ is the uniform limit of a sequence of locally constant maps.

Now, we consider one condition for $X$ under which the only continuous maps are the locally constant ones. First we prove,

**PROPOSITION 2.1.** *The following properties are equivalent for* $X$,

(1) *Every* $G_\delta$ *subset of* $X$ *is open.*

(2) *Every zero-set of* $X$ *is open.*

(3) *The intersection of a countable family of clopen subsets of* $X$ *is clopen.*

(4) *Every continuous map* $f : X \to K$ *is locally constant.*

(5) *Every continuous map* $f : X \to K$ *with a countable range is locally constant.*

*Proof.* (1) $\Leftrightarrow$ (2) is because $G_\delta$ is the same as zero-set ([Be], proposition 1). (1) $\Rightarrow$ (3) is obvious. As for (3) $\Rightarrow$ (4) let $f \in C(X)$ and let $f(x) = p$. Then,

$$\{y \in X : f(x) = p\} = \bigcap_n \{y \in X : |f(x)-p| \leq 1/n\}$$

is clopen.

(4) $\Rightarrow$ (5) is obvious. Finally let us prove (5) $\Rightarrow$ (1).

Let $G = \bigcap_n U_n$ be a $G_\delta$ subset of $X$ and let $a \in G$. Set $W_0 := X$, and for $n \geq 1$, let $W_n$ be a clopen subset of $X$ such that

$$a \in W_n \subset W_0 \cap W_1 \cap \ldots \cap W_{n-1} \cap U_n.$$

Let $\lambda$ be an element of $K$ such that $0 < |\lambda| < 1$. Define $f : X \to K$ by $f(x) = 0$ if $x \in \bigcap_n W_n$ and $f(x) = \lambda^n$ if $x \in W_n - W_{n-1}$. It is easy to see that $f$ is continuous. By hypothesis $f$ is locally constant and so

$$\bigcap_n W_n = \{x \in X : f(x) = 0\} \subset G$$

is a neighborhood of $a$. Hence, $G$ is open.

Following [G] we call those spaces verifying one of the above equivalent conditions, $P$-spaces. Notice again that the above proposition fails for non zerodimensional spaces.

**PROPOSITION 2.2.** *The following are equivalent for $X$,*

(1) *$X$ is a $P$-space.*
(2) *Every ideal $I$ in $C(X)$ is a $z$-ideal (that is: if $z(f) = z(g)$ for some $g \in I$, then $f \in I$).*
(3) *Every ideal in $C(X)$ is an intersection of prime ideals.*
(4) *The ideals spanned by $f$ and $f^2$ respectively coincide for each $f \in C(X)$.*
(5) *Every ideal in $C(X)$ is an intersection of maximal ideals.*
(6) *Every principal ideal in $C(X)$ is generated by an idempotent.*
(7) *Every prime ideal in $C(X)$ is maximal.*

Before proving the theorem we introduce some extra terminology: For a given $p \in X$ set $O_p := \{f \in C(X)$: $z(f)$ is a neighborhood of $p\}$ and $M_p := \{f \in C(X)$: $f(p) = 0\}$. With this terminology $X$ is a $P$-space if and only if $M_p = O_p$ for all $p \in X$. Moreover $O_p$ and $M_p$ are both ideals of $C(X)$; $O_p$ is a $z$-ideal and $M_p$ is a maximal ideal containing $O_p$. Moreover,

**LEMMA 2.3.** *If $I$ is any ideal containing $O_p$, then $I$ is contained in $M_P$. In particular, if $I$ is also maximal, then $I = M_p$.*

*Proof.* First recall that for any ideal $M$ in $C(X)$ the following property holds ([Be], included in the proof of the proposition 3): If $f, g \in M$ then $z(f) \cap z(g) \neq \emptyset$.

Assume now the existence of an element $f \in I$ such that $p \notin z(f)$. Let $S$ be a clopen subset of $X$ such that $z(f) \subset S$ and $p \in X - S$. Then $k_S \in O_p \subset I$. However, $z(f) \cap z(k_S) = \emptyset$.

*Proof* (of (2.2)):

(1) $\Rightarrow$ (2). Let $I$ be an ideal in $C(X)$ and $f \in C(X)$ such that $z(f) = z(g)$ for some $g \in I$. Since by hypothesis $z(f)$ is clopen, the map $h : X \to K$ defined by $h(x) = f(x)/g(x)$ if $x \notin z(g)$ and $h(x) = 0$ otherwise is continuous. Hence $f = gh \in I$.

(2) $\Rightarrow$ (3). It is sufficient to prove that $I$ coincides with its radical $r(I) := \{f \in C(X)$: $f^n \in I$ for some $n > 0\}$. Indeed if $f^n \in I$ for some $n$, then $f \in I$ because $z(f^n) = z(f)$ and $I$ is a $z$-ideal.

(3) $\Rightarrow$ (4). It is enough to observe that the prime ideals containing $f$ are the same as those containing $f^2$.

(4) $\Rightarrow$ (5). Let $I$ be an ideal and let $T$ be the intersection of all the maximal ideals containing $I$. We shall first prove that $T$ is contained in $r(I)$: let $f \in T$, let $P$ be a prime ideal containing $I$ and let $M$ be a maximal ideal containing $P$. From (4), $f = f^2g$ for some $g \in C(X)$ and then $0 = f(fg-1) \in P$. Moreover $fg-1$ is not in $P$ because $P \subset M$ and $f \in M$. Hence $f \in P$.

Since $f \in r(I)$, there exists $n > 0$ such that $f^n \in I$. By hypothesis the ideal generated by $f$ coincides with the ideal generated by $f^n$ and so $f \in I$.

(5) $\Rightarrow$ (6). Let $I = (f)$ be the principal ideal generated by $f \in C(X)$. Since the maximal ideals containing $f$ coincide with the maximal ideals containing $f^2$ then $(f) = (f^2)$ and hence there exists $g \in C(X)$ such that $f = f^2g$. Then $I$ coincides with the ideal generated by the idempotent $fg$.

(6) $\Rightarrow$ (7). Let $P$ be a prime ideal and $f \in C(X)-P$. By hypothesis (f) = (g) for some idempotent $g \in C(X)$. Then $g(g-1) \in P$ and so $g-1 \in P$. Thus $P+(f) = C(X)$ and $P$ is maximal.

(7) $\Rightarrow$ (1). Let $p \in X$. Since $O_p$ is a $z$-ideal it coincides with the intersection of all the prime (and hence maximal) ideals containing it. It follows from the lemma that $M_p = O_p$.

**Remarks**

**2.4.** If $K$ is such that $-1 \notin K^2$, then the (1)-(7) properties are also equivalent to

(8) *The ideals* $(f,g)$ *and* $(f^2+g^2)$ *coincide for all* $f,g \in C(X)$.

Indeed (8) $\Rightarrow$ (4). As for (1) $\Rightarrow$ (8) notice first that for any $f,g \in C(X)$, $z(f^2+g^2) = z(f) \cap z(g)$.

Define $h \in C(X)$ by $h(x) = f(x)/(f^2(x)+g^2(x))$ if $x \notin z(f^2+g^2)$ and $h(x) = 0$ otherwise. Then $f = (f^2+g^2)h \in (f^2+g^2)$. Thus $(f,g) = (f^2+g^2)$.

**2.5.** The implications (3) $\Rightarrow$ (4) $\Leftrightarrow$ (5) $\Leftrightarrow$ (6) $\Leftrightarrow$ (7) hold in any commutative ring. However, the ideal $(x_1, x_2, \ldots)/I$ is the only prime ideal in the ring $K[x_1, \ldots, x_n, \ldots]/I$ where $I = (x_1, x_2^2, \ldots, x_n^n, \ldots)$; so that (7) $\Rightarrow$ (3) does not hold in general.

The following result is the zerodimensional counterpart to the well known result saying that a topological space is a $P$-space if and only if its realcompactification is a $P$-space.

THEOREM **2.6.** *$X$ is a $P$-space if and only if $X^\delta$ is a $P$-space.*

*Proof.* First assume $X$ is a $P$-space and let $f : X^\delta \to K$ be a continuous map with a countable range. Then $f|X : X \to K$ is locally constant; hence there exists a clopen partition $\{B_n : n \in \mathbb{N}\}$ of $X$ such that $f$ is constant in $B_n$, say $f(B_n) = a_n$. Now, set $C_i := \overline{B}_i$, the closure in $X^\delta$, for each $i \in \mathbb{N}$. Then, $\{C_n : n \in \mathbb{N}\}$ is a clopen partition of $X^\delta$ and $f(C_n) = a_n$. Hence $f$ is locally constant.

Now assume $X^\delta$ to be a $P$-space and let $K$ be a locally compact field. If $f : X \to K$ is a continuous map, then there is a continuous extension $f^\delta : X^\delta \to K$ of $f$ to $X^\delta$ because $K$ is $\mathbb{N}$-compact. By hypothesis $f^\delta$ is locally constant and hence so is $f$.

$P$-spaces are almost discrete spaces. The previous result allows us to give an example of a non discrete $P$-space.

EXAMPLE **2.7.** Assume the existence of (one) measurable cardinal $m$ and let $X$ be any set with $\#X = m$ endowed with the discrete topology. According to the previous result $X^\delta$ is a $P$-space. Moreover $X^\delta$ is not discrete because it is $\mathbb{N}$-compact and its cardinal is measurable.

Finally the following corollary shows a certain duality between pseudocompact and $P$-spaces.

COROLLARY **2.8.** *If the valuation of $K$ is dense, then $BC_s(X)$ is spherically complete if and only if $X$ is a $P$-space and $K$ is spherically complete.*

*Proof.* It follows from [Vr2] (theorem 4.1) and the preceding theorem.

## REFERENCES


[Ba] G. Bachman, E. Beckenstein, L. Narici and S. Warner. *Rings of continuous functions with values in a topological field.* Trans. Amer. Math. Soc. **204** (1975), 91–112.

[Be] E. Beckenstein, G. Bachman and L. Narici. *Topological algebras of continuous functions over valued fields.* Studia Math. **48** (1973), 119–127.

[D] J. Dugundji. *Topology.* Allyn and Bacon, Boston, 1966.

[E] R. Engelking. *General Topology.* Helderman, Berlin, 1989.

[El] E. Ellis. *Extending continuous functions on zero-dimensional spaces.* Math. Ann. **186** (1970), 114–122.

[G] L. Gillman and M. Jerison. *Rings of continuous functions.* Springer-Verlag, 1960.

[M] J. Martínez-Maurica and S. Navarro. *p-adic Ascoli theorems.* Rev. Mat. Univ. Complut. Madrid **3** (1990), 19–27.

[N] P. Nyikos. *Review of* [Ba]. Mathematical Reviews **53** (1977) #6503.

[N2] P. Nyikos and H.C. Reichel. *On the structure of zerodimensional spaces.* Nederl. Akad. Wetensch. Proc. Ser. A **78** = Indag. Math. **37** (1975), 120–136.

[P] R.S. Pierce. *Rings of integer-valued continuous functions.* Trans. Amer. Math. Soc. **100** (1961), 371–394.

[Sch] W. Schikhof. *Non-archimedean Calculus.* Report 7812, Dept. Math., Catholic University, Nijmegen, The Netherlands (1978).

[Vr] A.C.M. van Rooij. *Non-archimedean functional analysis.* Marcel Dekker, New York, 1978.

[Vr2] A.C.M. van Rooij. *Notes on p-adic Banach spaces.* Report 7633, Dept. Math., Catholic University, Nijmegen, The Netherlands (1976).



J. Araujo
Departamento de Matemáticas, Universidad de Oviedo
Edificio Facultad de Ciencias
Calvo Sotelo s/n 33007, Oviedo, Spain

P. Fernández-Ferreirós, J. Martínez-Maurica
Departamento de Matemáticas
Facultad de Ciencias, Universidad de Cantabria
Av. de los Castros, 39071 Santander, Spain

# EXTENSION OF ISOMETRIES WITH VALUES IN NONARCHIMEDEAN FIELDS

José M. Bayod*

## 1. INTRODUCTION

Assume $K$ is a complete nonarchimedean nontrivially valued field, $A$ is a subset of $K$, and $f : A \longrightarrow K$ is an *isometric map*, or an *isometry*, meaning a (not necessarily onto) map for which

$$\forall x, y \in A, \quad |f(x) - f(y)| = |x - y|$$

holds. It is well-known, [2, Theorem 6], that when $K$ is spherically complete, the map $f$ can always be extended isometrically at least up to the set of so-called *strictly efficient* points of $K$ with respect to $A$, namely

$$c_K(A) := \{x \in K : \forall y \in K, \exists a \in A, \; d(a, x) < d(a, y)\}.$$

On the other hand, W. Schikhof has characterized the fact that all such isometries defined over a subset of $K$ can be extended to the whole $K$:

THEOREM [4, Theorems 75.4 and 76.4].- *The following conditions are equivalent:*

($\alpha$) *$K$ is spherically complete. Its residue class field is finite.*

($\beta$) *Each isometry $K \longrightarrow K$ is surjective.*

($\gamma$) *Each $K$-valued isometry defined on a subset of $K$ can be extended to an isometry $K \longrightarrow K$.*

In this note we shall give new extension results for the case when the residue class field of $K$ is infinite.

Given any valued field $K$, we will denote by $\overline{K}$ the residue class field of $K$.

* Research partially supported by DGICYT (grant PS87-0094)

## 2. RESULTS

**LEMMA 1.** *Assume that the subset $A \subset K$ is finite and the residue class field $\overline{K}$ is infinite. Then for each point $z \in K \setminus A$, any isometry $f : A \longrightarrow K$ can be extended to an isometry $A \cup \{z\} \longrightarrow K$.*

*Proof.* Let $A = \{a_1, \ldots, a_n\}$, and assume that the members of $A$ are labeled in such a way that

$$|z - a_1| = \ldots = |z - a_p| < |z - a_{p+1}| \leq \ldots \leq |z - a_n| \qquad (1 \leq p \leq n).$$

Then for each $i \leq p$, $\left|\dfrac{f(a_i) - f(a_1)}{z - a_1}\right| \leq 1$, and since $\overline{K}$ is infinite, there exists a $\lambda$ in $K$ such that $|\lambda| = 1$ and

$$\forall i = 1, \ldots, p, \qquad \left|\lambda - \frac{f(a_i) - f(a_1)}{z - a_1}\right| = 1.$$

Define $\overline{f}(z) := \lambda(z - a_1) + f(a_1)$, and $\overline{f}\,|_A = f$. We claim that

$$\forall i = 1, \ldots, n, \qquad |\overline{f}(z) - f(a_i)| = |z - a_i| :$$

- if $1 \leq i \leq p$,

$$|\overline{f}(z) - f(a_i)| = |z - a_1| \left|\lambda - \frac{f(a_i) - f(a_1)}{z - a_1}\right| = |z - a_1| = |z - a_i|;$$

- if $p < i \leq n$, then $|z - a_i| > |z - a_1|$, so $|a_i - a_1| = |z - a_i|$; moreover,

$$|f(a_i) - f(a_1)| = |a_i - a_1| = |z - a_i| > |z - a_i| = |\overline{f}(z) - f(a_1)|,$$

  therefore $|\overline{f}(z) - f(a_i)| = |z - a_i|$.

**THEOREM 1.** *Assume that $K$ is complete and its residue class field $\overline{K}$ is infinite. If $A \subset X \subset K$, $A$ is finite, and $X$ is separable, then any isometry $f : A \longrightarrow K$ has an isometric extension $X \longrightarrow K$. In particular, when $K$ is separable, every $K$-valued isometry defined on a finite subset of $K$ can extended to the whole $K$.*

*Proof.* Let $\{x_n : n \in \mathbb{N}\} \subset X$ be dense in $X$. Applying last lemma, we prove by induction that for each natural number $n$ there is an isometry

$$f_n : A \cup \{x_1, \ldots, x_n\} \longrightarrow K$$

which extends $f_{n-1}$ (call $f_0 := f$). Then define $g : A \cup \{x_n\} \longrightarrow K$ by

$$g(x_n) := f_n(x_n) \quad \forall n \in \mathbb{N}$$
$$g(a) := f(a) \quad \forall a \in A,$$

which is also isometric. Since $g$ is uniformly continuous and $K$ is complete, $g$ has a continuous extension to $\overline{A \cup \{x_n\}} = X$, and this extension is obviously isometric.

According to W. Schikhof's results quoted in the introduction, in case $\overline{K}$ is finite but $K$ is *not* spherically complete, there exist surjective isometries from certain subsets of $K$ onto $K$, making it impossible the existence of an isometric extension of them to the whole $K$; however, one can *make more room* for the extension by enlarging the field: if $\check{K}$ is the spherical completion of $K$ (notation borrowed from [6]), since $\overline{\check{K}} = \overline{K}$ is still finite, any isometry $A \longrightarrow K$ (with $A \subset K$) can be extended to an isometry $\check{K} \longrightarrow \check{K}$.

What we plan to do next is obtaining a similar result, though via a very different construction, for the case when $\overline{K}$ is infinite (and $K$ is separable).

Consider a nonstandard $\aleph_1$-saturated model of $K$, ${}^*K$, such as, for instance, an $\aleph_1$-saturated ultrapower of $K$. For notations and background on the properties of the nonstandard models we are going to use, we refer to the first chapter of the books [5] and [3]. The operations of addition and multiplication, and the absolute value of $K$ are extended in a canonical way up to ${}^*K$, and ${}^*K$ and $K$ are related by the Transfer Principle (all properties of the standard model can be transferred, when properly stated, to the nonstandard model), so that ${}^*K$ is a larger field (we are assuming that the valuation of $K$ is not trivial) endowed with a generalized valuation that takes values in the set of hyperreal numbers, ${}^*\mathbb{R}$. Call

$$\text{fin}\,{}^*K := \{x \in {}^*K : |x| \text{ is finite}\}$$
$$\text{inf}\,{}^*K := \{x \in {}^*K : |x| \text{ is infinitesimal}\}$$
$$\widehat{K} := \text{fin}\,{}^*K / \text{inf}\,{}^*K,$$

and endow the field $\widehat{K}$ (the *infinitesimal hull* or *nonstandard hull* of $K$), that contains a copy of $K$ ($x \mapsto \widehat{x}$) with a rank one valuation via

$$\widehat{K} \longrightarrow \mathbb{R}$$
$$\widehat{x} \longmapsto |\widehat{x}| := \text{st}\,|x|,$$

where st is the *standard part* map, that sends every finite hyperreal number onto the only real number infinitely close to it. Then ([1, Theorem 2]) $\widehat{K}$ is spherically complete; the value group of $\widehat{K}$ is the closure in $(0, +\infty)$ of the value group of $K$; $\overline{\widehat{K}} \supset {}^*(\overline{K})$ and these fields are equal when the valuation is discrete; finally, ${}^*(\overline{K}) \supset \overline{K}$, and they coincide if and only if $\overline{K}$ is finite. The infinitesimal hull of $K$ is the same as the infinitesimal hull of the completion of $K$.

**THEOREM 2.** *Let $K$ be separable. Then there exists a valued field $L$ greater than $K$ such that for any subset $A \subset K$ and any isometry $f : A \longrightarrow K$, there is an isometric extension of $f$ of the form $L \longrightarrow L$.*

*Proof.* As remarked above, when $\overline{K}$ is finite we can take as $L$ the spherical completion of $K$. Assume then that $\overline{K}$ is infinite.

Let $L$ be a nonstandard hull of $K$. Then $\overline{L} \supset {}^*K$ is also infinite, and we can assume, without loss of generality, that $K$ was complete, by the last remark preceeding this theorem. Then theorem 1 allows us to state that *every $K$-valued isometry defined on any finite subset of $K$ can be extended to an isometry $K \longrightarrow K$.* And the Transfer Principle applied to the sentence in italics implies that *every ${}^*K$-valued internal isometry defined on any ${}^*$-finite subset of ${}^*K$ can be extended to an (internal) isometry ${}^*K \longrightarrow {}^*K$.*

On the other hand, since the nonstandard model is $\aleph_1$-saturated, every countable standard set is contained in a ${}^*$-finite set; call $X$ a countable dense subset of $K$ such that $X \cap A$ is dense in $A$, and $F$ a ${}^*$-finite subset of ${}^*X \cap {}^*A$ that contains $X \cap A$. Consider the nonstandard extension of $f$ to a map ${}^*A \longrightarrow {}^*K$, and call $g$ its restriction to $F$, which is an internal isometry:

$$\forall x, y \in F, \quad |g(x) - g(y)| = |f(x) - f(y)| = |x - y|.$$

Therefore, $g$ can be extended to an isometry $h : {}^*K \longrightarrow {}^*K$. Notice that $g$ takes finite points of ${}^*K$ into finite points: take a fixed standard $a \in X \cap A$, then for any $x \in \operatorname{fin} {}^*K$,

$$|h(x) - h(a)| = |x - a| \leq |x| + |a|,$$

so that $|h(x)|$ is also finite. And notice also that an infinitesimal perturbation in fin ${}^*K$ produces only an infinitesimal perturbation in its images under $h$:

$$\forall x, y \in \operatorname{fin} {}^*K, \quad \widehat{x} = \widehat{y} \Rightarrow |h(x) - h(y)| = |x - y| \approx 0 \Rightarrow \widehat{h(x)} = \widehat{h(y)}.$$

Then the map $h$ induces a map in the quotient fin ${}^*K/\inf {}^*K$:

$$\overline{f}:\widehat{K}\longrightarrow\widehat{K}$$
$$\widehat{x}\longmapsto\widehat{h(x)}$$

which is isometric: for any $\widehat{x},\widehat{y}\in\widehat{K}$,

$$|\overline{f}(\widehat{x})-\overline{f}(\widehat{y})|=|\widehat{h(x)}-\widehat{h(y)}|=\mathrm{st}\,|h(x)-h(y)|=\mathrm{st}\,|x-y|=|\widehat{x}-\widehat{y}|,$$

and extends the original $f$ (regarded as a map between the copies of $A$ and $K$ contained in $\widehat{K}$ through $x\mapsto\widehat{x}$): if $a\in X\cap A$,

$$\overline{f}(\widehat{a})=\widehat{h(a)}=\widehat{g(a)}=\widehat{f(a)},$$

and $X\cap A$ is dense in $A$.

When the valuation is discrete, nonstandard hulls (or, for that matter, ultrapowers) of separable fields enjoy an additional extension property which is stronger than the one proved in Theorem 1, namely, there are isometric extensions not only from isometries defined over finite subsets, but also from all isometries defined over separable subsets:

**THEOREM 3.** *Let $K$ be separable and discretely valued. If $A\subset\widehat{K}$ is separable and $f:A\longrightarrow\widehat{K}$ is an isometry, then $f$ can be extended to an isometry $\widehat{K}\longrightarrow\widehat{K}$.*

*Proof.* If $\overline{K}$ is finite, then $\overline{\widehat{K}}=\overline{K}$ and $\widehat{K}$ is in fact the completion of $K$, which is spherically complete. Then the result follows from [4, Theorem 76.4]. Assume, then that $\overline{K}$ is infinite. In this case, the valuation of $\widehat{K}$ is also discrete (its value group is the same as the value group of $K$), and $\overline{\widehat{K}}={}^*(\overline{K})$ is also infinite.

**Case I:** $A$ is finite. Let $n$ be the number of points in the set $A$, $A=\{\widehat{a}_1,\ldots,\widehat{a}_n\}$. Define

$$g:\{a_1,\ldots,a_n\}\longrightarrow{}^*K$$
$$a_i\longmapsto g(a_i)$$

by choosing as $g(a_i)$ a representative in fin ${}^*K$ of $f(\widehat{a}_i)\in\widehat{K}$. The set $\{a,\ldots,a_n\}\subset{}^*K$ is ${}^*$-finite, and the function $g$ is internal. Notice also that $g$ is isometric, since

$$\forall i,j\in\mathbf{N},\ i\neq j\ \Rightarrow\ |g(a_i)-g(a_j)|\approx|f(\widehat{a}_i)-f(\widehat{a}_j)|=|\widehat{a}_i-\widehat{a}_j|\approx|a_i-a_j|,$$

and this forces the equality because $K$ is strictly valued and we are assuming that $\widehat{a_i} \neq \widehat{a_j}$ whenever $i \neq j$. Therefore, by transfer of Theorem 1, there exists an isometric extension $h : {}^*K \longrightarrow {}^*K$. Then

$$\overline{f} : \widehat{K} \longrightarrow \widehat{K}$$
$$\widehat{x} \longmapsto \overline{f}(\widehat{x}) := \widehat{h(x)}$$

is well defined and is an isometric extension of $f$:

- If $x \in \text{fin}\,{}^*K$, then $|x - a_1|$ is finite, $|h(x) - h(a_1)| = |x - a_1|$ and $h(x) \in \text{fin}\,{}^*K$:

$$|h(x)| \leq |x - a_1| + |h(a_1)|$$

- If $x, y \in \text{fin}\,{}^*K$ and $x \approx y$, then $|h(x) - h(y)| = |x - y| \approx 0$; threfore, $\widehat{h(x)} = \widehat{h(y)}$.
- If $x, y \in \text{fin}\,{}^*K$,

$$|\overline{f}(\widehat{x}) - \overline{f}(\widehat{y})| = |\widehat{h(x)} - \widehat{h(y)}| \approx |h(x) - h(y)| = |x - y| \approx |\widehat{x} - \widehat{y}|,$$

and then the first and last terms are equal, since they are both standard real numbers.

**Case II:** $A$ is infinite. Let $\{\widehat{a_i} : i \in \mathbf{N}\}$ be a dense subset of $A$, assuming $\widehat{a_i} \neq \widehat{a_j}$ if $i \neq j$. By choosing as the image of $a_i$ any representative of $f(\widehat{a_i})$, the isometry $f$ induces a map $g_0 : \{a_i : i \in \mathbf{N}\} \longrightarrow {}^*K$ such that

$$\forall i, j \in \mathbf{N},\ i \neq j \Rightarrow |g_0(a_i) - g_0(a_j)| \approx |a_i - a_j|$$

and again this forces the exact equality, as in Case I.

Since our model is $\aleph_1$-saturated, the set $\{a_i : i \in \mathbf{N}\}$ is contained in some internal set $\{a_n : n \in {}^*\mathbf{N}\} \subset {}^*K$, and $g_0$ can be extended to an internal map $g : \{a_n : n \in {}^*\mathbf{N}\} \longrightarrow {}^*K$ (Cf. [3, Theorem 8.6]). This extension to $\{a_n : n \in {}^*\mathbf{N}\}$ need not be isometric, but since it is internal, the set

$$N_0 := \{n \in {}^*\mathbf{N} : \forall i, j \in {}^*\mathbf{N},\ i.j \leq n \ \Rightarrow\ |g(a_i) - g(a_j)| = |a_i - a_j|\}$$

is also internal, and $N_0 \supset \mathbf{N}$ because $g_0$ is isometric; taking into account taht $\mathbf{N}$ itself is not internal, we conclude that $N_0$ must contain some infinite hypernatural number.

Thus, for some $H \in {}^*\mathbf{N} \backslash \mathbf{N}$, the restriction $g\,|_F$, where $F := \{a_n : n \in {}^*\mathbf{N},\ 1 \leq n \leq H\}$, is an internal isometry. Now, $F$ is a $^*$-finite subset of ${}^*K$, so by Theorem 1 (transferred), we know that $g\,|_F$ can be extended to an isometry $h : {}^*K \longrightarrow {}^*K$. Then we define $\overline{f} : \widehat{K} \longrightarrow \widehat{K}$ by $\overline{f}(\widehat{x}) := \widehat{h(x)}$ and, reasonning like at the end of Case I, we prove that $\overline{f}$ is well defined, isometric, and that $\overline{f}$ extends the restriction of $f$ to the dense subset of $A$, $\{\widehat{a}_i i \in \mathbb{N}\}$. By uniqueness of extensions to the closure, $\overline{f}$ has to agree with $f$ on the whole $A$.

## REFERENCES


[1] José M. Bayod. *Spherical completeness with infinitesimals.* Rev. Mat. Hisp.-Amer. **40** (1980), 74–85.

[2] José M. Bayod and J. Martínez-Maurica. *Ultrametrically tight extensions.* In *Proceedings of the Conference on p-adic analysis (Houthalen, 1986)*, edited by N. De Grande-De Kimpe and L. van Hamme, Vrije Universiteit Brussel, Brussels (1987), 1–10.

[3] Albert E. Hurd and Peter A. Loeb. *An introduction to nonstandard real anlysis.* Academic Press, Orlando, 1985.

[4] W.H. Schikhof. *Ultrametric calculus.* Cambridge University Press, Cambridge, England, 1984.

[5] Keith D. Stroyan and José M. Bayod. *Foundations of infinitesimal stochastic analysis.* North-Holland, Amsterdam, 1986.

[6] A.C.M. van Rooij. *Non-archimedean functional analysis.* Marcel Dekker, New York, 1978.



Departamento de Matemáticas,
Facultad de Ciencias, Universidad de Cantabria,
Av. de los Castros,
39071 Santander, Spain

# WEAK $c'$-COMPACTNESS IN (STRONGLY) POLAR BANACH SPACES OVER A NON-ARCHIMEDEAN, DENSELY VALUED FIELD

Sabine Borrey

## 0. PRELIMINARIES

**0.1. PRELIMINARIES.** Unless stated otherwise, $K$ will be a non-archimedean (n.a.), complete valued field with non-trivial and ***dense*** valuation $|\cdot|$.
We put $B(0,1) = \{\lambda \in K \mid |\lambda| \leq 1\}$ and $B(0,1^-) = \{\lambda \in K \mid |\lambda| < 1\}$ and denote the residue class field of $K$ by $k$ and its value group by $|K^*|$.
Unless otherwise specified, $E$ will be a Banach space with norm $\|\cdot\|$. We write $(E, \|\cdot\|)$.
Throughout, $A \subset E$ is an ***absolutely convex*** (a.c.), ***infinite dimensional*** subset of $E$.
We denote by $E'$ the topological dual space of $E$ and by $\sigma(E, E')$ the weak topology on $E$. We will assume that $E' \neq \{0\}$.
For $S \subset E$, we denote by co$S$ the a.c. hull of $S$, by $\overline{\text{co}}S$ the closure of co$S$ and by $[S]$ the linear hull of $S$.
We consider the definitions of compactoidity, $c$-compactness, ($t$-)orthogonality and ($t$-) orthogonal bases in $(E, \|\cdot\|)$ to be widely known. (If not, we refer to [9] and [10].)

**0.2. ABSTRACT.** In this paper we give some characterizations of a.c., weakly $c'$-compact sets in (strongly) polar spaces and of a.c. weakly $c'$-compact sets that are compactoid. A topological vector space related to such sets is studied in order to give these characterizations. (Paragraphs **3**, **4** and **5**.)

## 1. POLAR AND STRONGLY POLAR SPACES

**Remark 1.1.** Throughout **1**, $E$ is a Hausdorff locally convex space.

**DEFINITION 1.2.** ([7]). A seminorm $p$ on $E$ is a polar seminorm if $p = \sup\{|f| \mid f \in E^*, |f| \leq p\}$.

**DEFINITION 1.3.** ([7]). Let $B \subset E$. Put $B^0 = \{f \in E' \mid |f(x)| \leq 1 \; \forall x \in B\}$ and $B^{00} = \{x \in E \mid |f(x)| \leq 1 \; \forall f \in B^0\}$. $B$ is called a polar set if $B = B^{00}$.

**PROPOSITION 1.4.** *We have the following:*

i) *Each polar set in E is closed and edged;*

ii) *A subset B of E is a polar set if and only if there exists a collection $\mathcal{F} \subset E'$ such that* $B = \bigcap_{f \in \mathcal{F}} \{x \in E \mid |f(x)| \leq 1\}$.

*Proof.* [7], proposition 3.4.

**DEFINITION 1.5.** ([7]).

1) $E$ is a strongly polar space if every continuous seminorm on $E$ is polar.

2) $E$ is a polar space if its topology can be defined by a family of polar seminorms.

**THEOREM 1.6.** *The following are equivalent:*

i) *E is strongly polar;*

ii) *For each linear subspace D, for each continuous seminorm p, for each $\varepsilon > 0$, for each $f \in D'$ with $|f| \leq p$ on D, there is a continuous extension $\overline{f}$ of f, such that* $|\overline{f}| \leq (1+\varepsilon)p$ *on E.* (Hahn-Banach property).

*Proof.* [7], theorem 4.2.

**PROPOSITION 1.7.** $A^{00} = (\overline{A}^{\sigma})^e$.

*Proof.* [7], proposition 4.10.

**COROLLARY 1.8.** *Let E be strongly polar. Then we have the following:*

*i)* $(\overline{A})^e = A^{00}$;

*ii) If A is closed (for the initial topology!), then* $A \subset \overline{A}^{\sigma} \subset A^e$.

*Proof.* [7], remark following proposition 4.10.

## 2. ON SEMINORMS ON $E$ AND (WEAKLY) $c'$-COMPACT SUBSETS IN A LOCALLY CONVEX SPACE $E$.

**Remark 2.1.** Throughout **2**, $E$ is a Hausdorff locally convex space.

**DEFINITION 2.2.** ([4]). Let $A$ be closed. We put $A^i = \bigcup_{\lambda \in B(0,1^-)} \lambda A$ and $\partial A = A \backslash \overline{A^i}$ where $\overline{A^i}$ is the closure of $A^i$ in $E$ (Note that $\partial A$ may be empty.) we denote by

$$\pi : A \to A/_{\overline{A^i}} : x \to \overline{x},$$

and by

$$\pi_1 : B(0,1) \longrightarrow k : \lambda \to \overline{\lambda}$$

the canonical surjections.
Then $V = A/_{\overline{A^i}}$ is a $k$-vector space.

**DEFINITION 2.3.** ([6] and [8]).

i) $A$ is called $c'$-compact, if for each continuous seminorm $p$ on $E$, there exists $x \in A$ such that $p(x) = \max_{y \in A} p(y)$.

ii) $A$ is called weakly $c'$-compact, if for each $f \in E'$, there exists $x \in A$ such that $|f(x)| = \max_{y \in A} |f(y)|$.

**PROPOSITION 2.4.**

i) *If $A$ is $c'$-compact, then $A$ is bounded.*

ii) *If $E$ is a polar space and if $A$ is weakly $c'$-compact, then $A$ is bounded.*

*Proof.* i) Obvious.

ii) [7], corollary 7.7.

**DEFINITION 2.5.** ([8]). $A$ is a pure compactoid if for each zero-neighbourhood $U$, there exists $F \subset A$, finite, such that $A \subset U + \text{co}F$.

**PROPOSITION 2.6.** *$A$ is a pure compactoid if and only if $A$ is $c'$-compact.*

*Proof.* In [6], theorem 2.7. ($\alpha$) $\Rightarrow$ ($\beta$), it is proved that a weakly $c'$-compact subset of $E$ is $c'$-compact for the weak topology (the proof works in any locally convex space). This is important, because when comparing definitions 2.3.i) and 2.3.ii), $c'$-compactness for the weak topology and weak $c'$-compactness seem to be different properties.
Finally, it is proved in [8], theorem 3.3 that in any locally convex space, $A$ is a pure compactoid if and only if $A$ is $c'$-compact.

## 3. A TOPOLOGICAL VECTOR SPACE RELATED TO $A$

**Remark 3.1.** Throughout **3**, $A$ is closed.

**CONSTRUCTION 3.2.**

i) With the notations of 2.2, the formula $\|\pi(x)\| = \text{dist}(x, \overline{A^i})$, defines a norm on $V$. This norm induces a topology on $V$ which we will use in the sequel.

ii) The topology on $A$ is the topology induced by the norm on $E$.

iii) It is clear that now $\pi : (A, \|\cdot\|) \longrightarrow (V, \|\cdot\|) : x \mapsto \overline{x}$, is continuous.

iv) Notations and definitions introduced in 3.2, are valid throughout the rest of the text.

**PROPOSITION 3.3.** $(V, \|\cdot\|)$ *is complete.*

*Proof.* Straightforward.

**PROPOSITION 3.4.** *If* $A = \overline{\text{co}}X$, *then* $\overline{[\pi(X)]} = V$.

*Proof.* As $\pi$ is continuous $V = \pi(\overline{\text{co}}X) \subset \overline{\pi(\text{co}X)} = \overline{[\pi(X)]} \subset V$.

**PROPOSITION 3.5.** *For each* $t \in (0,1)$, $(V, \|\cdot\|)$ *has a (t-)orthogonal base.*

*Proof.* [3], corollary 2.5.

**Remark 3.6.** In view of proposition 3.4, we are going to consider ($t$-)orthogonality in $(V, \|\cdot\|)$ versus ($t'$-)orthogonality in $(E, \|\cdot\|)$ $(t' \in (0,1))$.

**PROPOSITION 3.7.** *Let* $t \in (0,1)$, *let* $(e_i)_{i \in I}$ *be an infinite (t-)orthogonal family in* $(E, \|\cdot\|)$ *and suppose that there is no restriction on the valuation on* $K$. *Then we have the following:*

i) $(e_i)_{i \in I}$ *is a (t-)orthogonal base of* $\overline{[\{e_i \mid i \in I\}]}$ *and* $\overline{[\{e_i \mid i \in I\}]} = \{\sum_{i \in I} \lambda_i e_i \mid \forall i \in I : \lambda_i \in K$ *and* $\lambda_i . e_i \rightarrow 0\}$.

ii) $\overline{\text{co}}(e_i \mid i \in I) = \{\sum_{i \in I} \lambda_i e_i \mid \forall i \in I : \lambda_i \in B(0,1)$ *and* $\lambda_i . e_i \rightarrow 0\}$.

*Proof.* i) The proof is given for orthogonal families in [10], lemma 5.1. With some minor modifications, the proof can be used for ($t$)-orthogonal families. Both proofs also work in case the valuation on $K$ is trivial.

ii) Clearly, $\text{co}(e_i \mid i \in I) \subset \{\sum_{i \in I} \lambda_i.e_i \mid \forall i \in I : \lambda_i \in B(0,1) \text{ and } \lambda_i.e_i \to 0\} \subset \overline{\text{co}}(e_i \mid i \in I)$, and it is not hard to prove that $\{\sum_{i \in I} \lambda_i.e_i \mid \forall i \in I : \lambda_i \in B(0,1) \text{ and } \lambda_i.e_i \to 0\}$ is closed in $\overline{[\{e_i \mid i \in I\}]}$ and thus in $E$.

**PROPOSITION 3.8.** *Let $t \in (0,1)$ and let $(e_i)_{i \in I}$ be an infinite (t-)orthogonal family in $(E, \|\cdot\|)$. Put $A = \overline{\text{co}}(e_i \mid i \in I)$ and let $x = \sum_{i \in I} \lambda_i.e_i \in A$. Then the following are equivalent:*

i) *There exists $i_0 \in I$, such that $|\lambda_{i_0}| = 1$;*
ii) *$x \in \partial A$.*

*Proof.* i) $\Rightarrow$ ii). It is easy to see that $\overline{[A]} = \overline{[\{e_i \mid i \in I\}]}$.
Define $p : \overline{[A]} \to \mathbb{R}_+ : y = \sum_{i \in I} \mu_i.e_i \to |\mu_{i_0}|$. It is easy to see that $p$ is a seminorm on $\overline{[A]}$. Then for each $y \in \overline{[A]}$ with $y = \sum_{i \in I} \mu_i.e_i$, we have the following:

$$p(y) = |\mu_{i_0}| = \frac{1}{\|e_{i_0}\|}\|\mu_{i_0}.e_{i_0}\| \le \frac{1}{\|e_{i_0}\|}\max_{i \in I}\|\mu_i.e_i\| \le \frac{1}{t.\|e_{i_0}\|}\|y\|.$$

Hence, $p$ is continuous.
According to [8], proof of proposition 1.1, there exists a continuous seminorm $\overline{p}$ on $E$ such that $\overline{p}|_{\overline{[A]}} = p$.
Now, $1 = \overline{p}(x) \le \max_{y \in A} \overline{p}(y) = \max_{y \in A} p(y) \le \max_{i \in I} p(\mu_i.e_i) \le \max_{i \in I} p(e_i) = p(e_{i_0}) = 1$.
Hence, $\max_{y \in A} \overline{p}(y) = \overline{p}(x)$. It follows from [2], proposition 1.5 that $x \in \partial A$.

ii) $\Rightarrow$ i). $x \in \partial A = A \backslash \overline{A^i}$. There exists a continuous seminorm $p$ on $E$, $p \ne 0$ on $A$, such that $p(x) = \max_{y \in A} p(y)$ ([2], proposition 1.5.). But $p(x) \le \max_{i \in I} |\lambda_i|.p(e_i) \le p(x)$. Hence, there exists $i_0 \in I$ such that $|\lambda_{i_0}|.p(e_{i_0}) = p(x)$. Then obviously, $|\lambda_{i_0}| = 1$, because if $|\lambda_{i_0}| < 1$, then $p(e_{i_0}) > p(x)$, a contradiction.

**PROPOSITION 3.9.** *Let $t \in (0,1)$ and let $(e_i)_{i \in I}$ be an infinite (t-)orthogonal family in $(E, \|\cdot\|)$ such that $A = \overline{\text{co}}(e_i \mid i \in I)$. Then $\pi(e_i)_{i \in I}$ is a (t-)orthogonal base of $(V, \|\cdot\|)$.*

*Proof.*

1) $V = \overline{[\{\pi(e_i) \mid i \in I\}]}$.
   *Proof.* Proposition 3.4.
2) $(\pi(e_i))_{i \in I}$ is a $(t$-)orthogonal family in $(V, \|\cdot\|)$.

*Proof.* Let $J \subset I$ be finite and consider $\sum_{j\in J} \overline{\lambda_j}.\pi(e_j)$, where for each $j \in J : \overline{\lambda_j} \neq 0$.
We have to calculate $\| \sum_{j\in J} \overline{\lambda_j}.\pi(e_j)\|$.
To do so, we consider $x = \sum_{j\in J} \lambda_j.e_j$ and $y = \sum_{i\in I} v_i.e_i \in \overline{A^i}$ (where for each $i \in I : v_i \in B(0,1^-)$ and $\lim_{i\in I} v_i.e_i = 0$, according to proposition 3.8.) Clearly, we have to calculate $\|\pi(x)\|$.

$$\begin{aligned}\|x-y\| &= \|\sum_{j\in J}(\lambda_j - v_j).e_j - \sum_{j\in I\setminus J} v_j.e_j\| \\ &\geq (t).\max\{\max_{j\in J}|\lambda_j - v_j|.\|e_j\|, \max_{j\in I\setminus J}|v_j|.\|e_j\|\} \\ &\geq (t)\max_{j\in J}|\lambda_j - v_j|.\|e_j\| \geq (t).\max_{j\in J}|\lambda_j|.\|e_j\|.\end{aligned}$$

Hence, for each $y \in \overline{A^i}$, we have that $\|x-y\| \geq (t).\max_{j\in J}|\overline{\lambda_j}|.\|\pi(e_j)\|$.
As a consequence, $\| \sum_{j\in J} \overline{\lambda_j}.\pi(e_j)\| = \|\pi(x)\| = \inf_{y\in \overline{A^i}} \|x-y\| \geq (t).\max_{j\in J}|\overline{\lambda_j}|.\|\pi(e_j)\|$.
This means that $(\pi(e_i))_{i\in I}$ is a $(t$-)orthogonal family in $(V, \|\cdot\|)$.

3) From 1), 2) and proposition 3.7, we deduce that $(\pi(e_i))_{i\in I}$ is a $(t$-)orthogonal base of $(V, \|\cdot\|)$.

**CONSTRUCTION 3.10.** Let $t \in (0,1)$ and let $(s_i)_{i\in I}$ be a $(t)$-orthogonal family in $(V, \|\cdot\|)$. For each $i \in I$, we choose $e_i \in \partial A$ such that $\pi(e_i) = s_i$.

**LEMMA 3.11.** *With the notations of 3.10, let $(s_j)_{j\in J}$ be a subfamily of $(s_i)_{i\in I}$, such that $\inf_{j\in J}\|s_j\| = \alpha > 0$ and $\sup_{j\in J}\|e_j\| \leq \beta$. Then $(e_j)_{j\in J}$ is an $(\frac{\alpha}{\beta}t)$-orthogonal family in $(E, \|\cdot\|)$.*

*Proof.* Let $J'' \subset J$ be finite and consider $\sum_{j\in J''} \lambda_j.e_j$ where $\lambda_i \in K$, for each $i \in J''$.
Put $J_1 = \{i \in J'' \mid |\lambda_i| = \max_{j\in J''}|\lambda_j|\}$ and choose $j_0 \in J_1$ such that $|\lambda_{j_0}| = \max_{j\in J''}|\lambda_j|$.
We may assume that $\lambda_{j_0} \neq 0$. Then

$$\begin{aligned}\|\sum_{j\in J''}\lambda_i.e_i\| &= |\lambda_{j_0}|.\|\sum_{i\in J''}\frac{\lambda_i}{\lambda_{j_0}}.e_i\| \\ &\leq |\lambda_{j_0}|.\|\sum_{i\in J''}\overline{\left(\frac{\lambda_i}{\lambda_{j_0}}\right).s_i}\| \geq (t).|\lambda_{j_0}|.\max_{i\in J_1}\|s_i\| \\ &\geq \alpha.(t).|\lambda_{j_0}| = \alpha.(t).\max_{j\in J''}|\lambda_j| \geq \frac{\alpha}{\beta}.(t).\max_{j\in J''}|\lambda_i|.\|e_j\|.\end{aligned}$$

**PROPOSITION 3.12.** *Let $(s_i)_{i\in I}$ be an orthogonal family in $(V, \|\cdot\|)$. If for each $i \in I$, we can choose $e_i \in \partial A$, such that $\pi(e_i) = s_i$ and such that $\|e_i\| = \|s_i\|$, then $(e_i)_{i\in I}$ is an orthogonal family in $(E, \|\cdot\|)$.*

*Proof.* For $J \subset I$, finite, put $x = \sum_{j\in J} \lambda_j.e_j$ with $\lambda_j \in K$ for each $j \in J$. We will assume that $x \neq 0$. Put $L = \{j \in J \mid \|\lambda_j.e_j\| = \max_{i\in J} \|\lambda_i.e_i\| = \beta\}$ and choose $j_0 \in L$ such that $|\lambda_{j_0}| = \max_{i\in L} |\lambda_i|$ ($|\lambda_{j_0}| \neq 0$ as $x \neq 0$). Then

$$\begin{aligned}\beta &\geq \|\sum_{i\in L} \lambda_i.e_i\| = |\lambda_{j_0}|.\|\sum_{i\in L} \frac{\lambda_i}{\lambda_{j_0}}.e_i\| \geq |\lambda_{j_0}|.\|\sum_{i\in L} \overline{\left(\frac{\lambda_i}{\lambda_{j_0}}\right)} \cdot s_i\| \\ &\geq |\lambda_{j_0}|. \max_{\{i\in L \mid |\lambda_i| = |\lambda_{j_0}|\}} \|s_i\| = \max_{\{i\in L \mid |\lambda_i| = |\lambda_{j_0}|\}} |\lambda_{j_0}|.\|e_i\| \\ &= \max_{\{i\in L \mid |\lambda_i| = |\lambda_{j_0}|\}} \|\lambda_i.e_i\| = \beta.\end{aligned}$$

Hence, $\|x\| = \|\sum_{i\in L} \lambda_i.e_i + \sum_{i\notin L} \lambda_i.e_i\| = \|\sum_{i\in L} \lambda_i.e_i\| = \beta = \max_{i\in J} \|\lambda_i.e_i\|$.

**THEOREM 3.13.** *Let $D \subset V$ be a closed linear subspace of $V$ and let $t \in (0, 1)$. Then $D$ has a $(t)$-orthogonal complement (i.e. there exists a closed linear subspace $F$ of $V$ such that $D \cap F = \{0\}$, $D+F = V$ and if $x = d+f$ with $d \in D$, $f \in F$, then $\|x\| \geq t.\max\{\|d\|, \|f\|\}$).*

*Proof.* There exists a norm $p$ on $V$, such that for each $x \in V$, $t.\|x\| < p(x) \leq \|x\|$ and such that $p(V) \subset \{t^n | n \in \mathbb{Z}\} \cup \{0\}$. ([3], theorem 2.2.)

Let $S = (s_i)_{i\in I}$ be an orthogonal base of $(D, p)$. ([3], theorem 2.3.)

Put $\Omega = \{T \subset V \mid 0 \notin T, T \supset S$ and $T$ is an orthogonal family in $(V, p)\}$. A standard application of Zorn's lemma yields a maximal element $B$ of $\Omega$. We have the following:

1) $\overline{[B]} = V$.

   Indeed, suppose there exists $x \in V \backslash \overline{[B]}$. There exists $m \in \mathbb{Z}$ such that $t^m = \inf_{b\in\overline{[B]}} p(x-b)$. We can find $b_0 \in \overline{[B]}$, such that $p(x-b_0) < t^{m-1}$. Then $p(x-b_0) = \inf_{b\in\overline{[B]}} p(x-b)$, hence $x-b_0$ and $\overline{[B]}$ are orthogonal in $(V, p)$. But this is a contradiction with the maximality of $B$.

2) $B$ is an orthogonal base of $(V, p)$. (Proposition 3.7.i.)

3) $\overline{[B\backslash S]} + D = V$.

   Indeed, put $B = (b_j)_{j\in J}$ and $J_1 = \{j \in J \mid b_j \notin S\}$ and $J_2 = J\backslash J_1$.

   $x \in V \iff x = \sum_{j\in J} \lambda_j.b_j$ with $\lambda_j \in k$, hence $x = \sum_{j\in J_1} \lambda_j.b_j + \sum_{j\in J_2} \lambda_j.b_j$. It follows immediately that $x \in \overline{[B\backslash S]} + D$.

4) $\overline{[B\backslash S]} \cap D = \{0\}$.

This follows easily from the fact that $B$ is a base of $V$.

5) Let $x = d+f$ with $d \in D$ and $f \in F = \overline{[B\backslash S]}$.

Then $p(x) = \max\{p(d), p(f)\} \geq t.\max\{\|d\|, \|f\|\}$.

**DEFINITION 3.14.** $Y \subset V$ is called a generating subset of $V$ if and only if $\overline{[Y]} = V$.

**PROPOSITION 3.15.** *Let $Y \subset V$ be a generating subset of $V$ and let $T \subset A$ such that $\pi(T) = Y$. Then $A = \overline{\overline{A^i} + coT}$.*

*Proof.* We only have to prove that $A \subset \overline{\overline{A^i} + \text{co}T}$.

Let $x \in A$. As $V = \overline{[\pi(T)]}$, it follows that for each $\varepsilon > 0$, there exists $u \in \text{co}T$ such that $\|\pi(x) - \pi(u)\| = \text{dist}(x-u, \overline{A^i}) < \varepsilon$. Hence, for each $\varepsilon > 0$, there exist $u \in \text{co}T$ and $a \in \overline{A^i}$ such that $\|(x-u) - a\| < \varepsilon$. Obviously, $x \in \overline{\overline{A^i} + \text{co}T}$.

## 4. A CONNECTION BETWEEN *A* AND GENERATING SUBSETS OF *V*

**Remark 4.1.**

i) Throughout **4**, $(E, \|\cdot\|)$ is a *polar* Banach space.

ii) $A$ is *weakly $c'$-compact* and *polar*. (Hence, $A$ is closed. (Proposition 1.4.))

**CONSTRUCTION 4.2.** Let $Y \subset V$ be a generating subset of $V$ and choose $T \subset A$ such that $\pi(T) = Y$. Put $B = \text{co}T$.

Throughout the rest of **4**, the choice of $T$ will *not* be altered.

**PROPOSITION 4.3.** *$B$ is weakly $c'$-compact and $\max_{x \in A} |f(x)| = \max_{x \in B} |f(x)|$.*

*Proof.* Let $f \in E'$ such that $f \neq 0$ on $A$. Put $\alpha = \max_{x \in A} |f(x)|$.$A = \overline{\overline{A^i} + \text{co}T}$ (proposition 3.15) is weakly $c'$-compact, and thus $\overline{A^i} + B$ is weakly $c'$-compact ([8], proposition 1.2). Hence, there exists $z_0 \in \overline{A^i} + B$ such that $|f(z_0)| = \alpha$.

Also, there exist $a_0 \in \overline{A^i}$ and $u_0 \in B$ such that $z_0 = a_0 + u_0$. But $|f(a_0)| < \alpha$ ([2], lemma 1.2), hence $|f(u_0)| = |f(z_0) - f(a_0)| = \alpha$.

So, $\alpha = |f(u_0)| \leq \sup_{x \in B} |f(x)| \leq \max_{x \in A} |f(x)| = \alpha$ and $u_0 \in B$, hence $\alpha = \max_{x \in B} |f(x)|$.

**PROPOSITION 4.4.** $A = \left(\overline{B}^{\sigma}\right)^{e}$.

*Proof.* Obviously, $\left(\overline{B}^{\sigma}\right)^{e} \subset A$ because $\left(\overline{B}^{\sigma}\right)^{e} = B^{00}$ (proposition 1.7) and because $A$ is a polar set.
Suppose there exists $x \in A\backslash\left(\overline{B}^{\sigma}\right)^{e}$. $\left(\overline{B}^{\sigma}\right)^{e}$ is a polar set, hence there exists $\mathcal{F} \subset E'$ such that $\left(\overline{B}^{\sigma}\right)^{e} = \bigcap_{f\in\mathcal{F}} \{y \in E \mid |f(y)| \leq 1\}$. (Proposition 1.4).
As $x \notin \left(\overline{B}^{\sigma}\right)^{e}$, it follows that there exists $f \in \mathcal{F}$ such that $|f(x)| > 1$.
Then $1 < \max_{y\in A}|f(y)| = \max_{y\in B}|f(y)|$ (proposition 4.3.), but $\max_{y\in B}|f(y)| \leq 1$. This is obviously a contradiction.

**LEMMA 4.5.** *Let $D \subset E$ be a.c. Then $(D^e)^i = D^i$.*
*Proof.* We only have to prove that $(D^e)^i \subset D^i$.
Choose $x \in (D^e)^i$. Then there exist $\lambda \in B(0,1^-)$ and $x_\lambda \in D^e$ such that $x = \lambda.x_\lambda$.
Choose $\mu \in B(0,1^-)$ such that $|\mu| > |\lambda|$. As $x_\lambda \in D^e$, there exists $y_\lambda \in D$ such that

$$x_\lambda = \frac{\mu}{\lambda}.y_\lambda,$$

hence $x = \mu.y_\lambda$, so $x \in D^i$.

**PROPOSITION 4.6.** $A = \overline{B}^{\sigma}$.
*Proof.* $A = \left(\overline{B}^{\sigma}\right)^{e}$, hence $\overline{A^i} = \overline{\left(\left(\overline{B}^{\sigma}\right)^{e}\right)^{i}}$.
But, using lemma 4.5, we obtain: $\overline{\left(\left(\overline{B}^{\sigma}\right)^{e}\right)^{i}} = \overline{\left(\overline{B}^{\sigma}\right)^{i}}$, and also $\overline{\left(\overline{B}^{\sigma}\right)^{i}} \subset \overline{\overline{B}^{\sigma}} = \overline{B}^{\sigma}$.
Hence, $\overline{A^i} \subset \overline{B}^{\sigma}$. As a consequence, $A = \overline{\overline{A^i} + B} \subset \overline{\overline{A^i} + \overline{B}^{\sigma}} = \overline{\overline{B}^{\sigma}} = \overline{B}^{\sigma} \subset A$.

**COROLLARY 4.7.** *For $t \in (0,1)$, let $(s_i)_{i\in I}$ be a $(t)$-orthogonal base of $(V, \|\cdot\|)$ (proposition 3.5) and for each $i \in I$, choose $e_i \in \partial A$ such that $\pi(e_i) = s_i$. Then*
i) $A = \overline{\text{co}}^{\sigma}(e_i \mid i \in I)$.
ii) *If $E$ is strongly polar, then $A = \overline{\text{co}}(e_i \mid i \in I)$.*
*If the family $(s_i)_{i\in I}$ can be chosen orthogonal (i.e. if $(V, \|\cdot\|)$ has an orthogonal base) and if the family $(e_i)_{i\in I}$ is constructed as above, the same conclusions are valid.*
*Proof.* i) Obviously, $(s_i)_{i\in I}$ is a generating subset of $V$ and $\pi((e_i)_{i\in I}) = (s_i)_{i\in I}$. Now, use proposition 4.6 with $B = (s_i)_{i\in I}$.
ii) According to i) and corollary 1.8.ii), $A \subset \overline{\text{co}}^{\sigma}(e_i \mid i \in I) = (\overline{\text{co}}(e_i \mid i \in I))^e$, hence $A = (\overline{\text{co}}(e_i \mid i \in I))^e$. But, according to lemma 4.5, $A^i = ((\overline{\text{co}}(e_i \mid i \in I))^e)^i = (\overline{\text{co}}(e_i \mid i \in I))^i$, hence $\overline{A^i} \subset \overline{\text{co}}(e_i \mid i \in I)$. From proposition 3.15 we deduce that $A = \overline{\text{co}}(e_i \mid i \in I)$.

**LEMMA 4.8.** *Let $E$ be strongly polar and for $t \in (0,1)$, let $(s_i)_{i\in I}$ be a $(t)$-orthogonal base of $(V, \|\cdot\|)$. Then $\inf_{i\in I} \|s_i\| = 0$.*

*Proof.* Suppose $\inf_{i\in I} \|s_i\| > 0$. For each $i \in I$, choose $e_i \in \partial A$, such that $\pi(e_i) = s_i$. Clearly, $\inf_{i\in I} \|e_i\| > 0$ (as for each $i \in I$, $\|s_i\| \leq \|e_i\|$). Put $\alpha = \inf_{i\in I} \|s_i\|$ and $\beta = \sup_{i\in I} \|e_i\|$. (According to proposition 2.4, $A$ is bounded.) Then $(e_i)_{i\in I}$ is an $(\frac{\alpha}{\beta}t)$-orthogonal family in $(E, \|\cdot\|)$ (lemma 3.11). As $A = \overline{\text{co}}(e_i \mid i \in I)$ (corollary 4.7), it follows easily that $\overline{[A]} = \overline{[\{e_i \mid i \in I\}]}$.

Now $B = \{x \in \overline{[A]} \mid \|x\| \leq \frac{\alpha^2.t}{\beta}\} \subset A$.

Indeed, $x \in B$ if and only if $x = \sum_{i\in I} \lambda_i.e_i$ and $\|x\| \leq \frac{\alpha^2.t}{\beta}$.

Now, $\|x\| = \|\sum_{i\in I} \lambda_i.e_i\| \geq \frac{\alpha.t}{\beta}.\max_{i\in I} |\lambda_i|.\|e_i\| \geq \frac{\alpha^2.t}{\beta}.\max_{i\in I} |\lambda_i|$.

We obtain $\max_{i\in I} |\lambda_i| \leq 1$, hence $x \in \overline{\text{co}}(e_i \mid i \in I)$. (Proposition 3.7.ii).

It follows that $A$ is open in $\overline{[A]}$, but as $A$ is infinite-dimensional, this is a contradiction with [6], remark following lemma 2.2 and [6], lemma 2.4. (The proof of this lemma also works in strongly polar spaces.)

## 5. A CONNECTION BETWEEN $A$ AND $V$ IN STRONGLY POLAR BANACH SPACES

**Remark 5.1.**

i) Throughout 5, $(E, \|\cdot\|)$ is *strongly polar*.

ii) Unless stated otherwise, $A$ is *weakly c′-compact, closed* and *edged* (hence, polar), and *compactoid*.

**LEMMA 5.2.** *$A$ is a pure compactoid.*

*Proof.* Let $U$ be a zero neighbourhood for the initial topology. According to [7], theorem 5.12, there exists $W \in \sigma(E, E')$, zero neighbourhood, such that $W \cap A \subset U \cap A$. $A$ is weakly $c'$-compact, hence there exists $F \subset A$ finite, such that $A \subset W + \text{co}F$. (Proposition 2.6). Now, $(W + \text{co}F) \cap A = (W \cap A) + \text{co}F$, hence

$$A \subset (W \cap A) + \text{co}F \subset (U \cap A) + \text{co}F$$

and obviously, $(U \cap A) + \text{co}F \subset U + \text{co}F$.

**LEMMA 5.3.** *Let $U \subset V$ be open and a.c. Then there exists a finite subset $F$ of $V$, such that $V = U + [F]$. This means that $V$ is a* pure compactoid.

*Proof.* There exists $\varepsilon > 0$, such that $U_\varepsilon = \{x \in V \mid \|x\| < \varepsilon\} \subset U$.
Put $W_\varepsilon = \{y \in E \mid \|y\| < \varepsilon\}$. Clearly, $\pi(W_\varepsilon \cap A) \subset U_\varepsilon$. According to lemma 5.2, there exists $G \subset A$ finite, such that $A \subset W_\varepsilon + \mathrm{co}G$, hence $A = (W_\varepsilon \cap A) + \mathrm{co}G$. Put $F = \pi(G)$. Then $V = \pi(A) = \pi((W_\varepsilon \cap A) + \mathrm{co}G) \subset U_\varepsilon + \pi(\mathrm{co}G) \subset U_\varepsilon + [\pi(G)] \subset U + [F] \subset V$.

**LEMMA 5.4.** *For each $t \in (0,1)$, there exists a $(t)$-orthogonal sequence $(x_n)_{n\in\mathbb{N}_0}$ in $(V, \|\cdot\|)$ with $\lim_{n\to\infty} x_n = 0$ and $V = \overline{[\{x_n \mid n \in \mathbb{N}_0\}]}$.*

*Proof.* Let $t \in (0,1)$. There exists a $t$-orthogonal sequence $(a_n)_{n\in\mathbb{N}_0}$ with $\lim_{n\to\infty} a_n = 0$ and $A = \overline{\mathrm{co}}(a_n \mid n \in \mathbb{N}_0)$. ([8], theorem 2.7). Then $V = \overline{[\{\pi(a_n) \mid n \in \mathbb{N}_0\}]}$ (proposition 3.4). According to proposition 3.9, $(\pi(a_n))_{n\in\mathbb{N}_0}$ is a $(t)$-orthogonal base of $(V, \|\cdot\|)$.

**LEMMA 5.5.** *Let $p : V \to \mathbb{R}_+$ be a continuous seminorm, $p \neq 0$ on $V$. Then there exists $x \in V$ such that $p(x) = \max_{y\in V} p(y)$. This means that $V$ is $c'$-compact.*

*Proof.* According to lemma 5.4, there exists a $(t)$-orthogonal base $(x_n)_{n\in\mathbb{N}_0}$ of $(V, \|\cdot\|)$ such that $\lim_{n\to\infty} x_n = 0$ $(t \in (0,1))$.
$p$ is continuous, hence $\lim_{n\to\infty} p(x_n) = 0$. Clearly, for each $x \in V$, $p(x) \leq \sup_{n\in\mathbb{N}_0} p(x_n)$, hence $\sup_{y\in V} p(y) = \sup_{n\in\mathbb{N}_0} p(x_n) = \max_{n\in\mathbb{N}_0} p(x_n)$. Hence, there exists $n_0 \in \mathbb{N}_0$ such that $p(x_{n_0}) = \sup_{y\in V} p(y)$.

**Remark 5.6.**

1) Lemma 5.3 and lemma 5.5 show that although pure compactoidity and $c'$-compactness were defined in locally convex spaces over a n.a. non-trivially valued field, these definitions are not without meaning in Banach spaces over trivially valued fields. In the sequel, we will use the words *pure compactoid* and *$c'$-compact* in Banach spaces over trivially valued fields in the sense of definitions 2.3 and 2.5.

2) The notion weak $c'$-compactness however doesn't carry much weight in such spaces because, as the valuation on the base field is trivial, it is clear that all those spaces are weakly $c'$-compact.

3) It is not difficult to construct an example of an unbounded Banach space over a trivially valued field, hence, proposition 2.4 doesn't hold in such spaces.

**PROPOSITION 5.7.** *Let $(x_n)_{n\in\mathbb{N}_0}$ be a sequence in $V$ such that $\|x_1\| > \|x_2\| >$*

$\ldots > \|x_n\| > \ldots$ *(Note that according to lemma 4.8, there exist strictly decreasing sequences.) Then* $\lim_{n\to\infty} \|x_n\| = 0$.

*Proof.* Suppose that $\lim_{n\to\infty} \|x_n\| = \alpha > 0$. It is clear that $(x_n)_{n\in\mathbf{N}_0}$ is an orthogonal sequence in $(V, \|\cdot\|)$. For each $n \in \mathbf{N}_0$, choose $e_n \in \partial A$ such that $\pi(e_n) = x_n$ and $\|x_n\| \leq \|e_n\| < \|x_{n-1}\|$. It follows that $(\|e_n\|)_{n\in\mathbf{N}_0}$ is a strictly decreasing sequence. On the other hand, $(e_n)_{n\in\mathbf{N}_0}$ is an $(\frac{\alpha}{\|e_1\|})$-orthogonal sequence in $A$ (lemma 3.11). Clearly, $\lim_{n\to\infty} \|e_n\| \geq \alpha > 0$. This is a contradiction with the compactoidity of $A$ ([10]), theorem 4.37).

**THEOREM 5.8.** *Let $A \subset E$ be closed, edged and weakly $c'$-compact. Then the following are equivalent:*

i) *$A$ is compactoid;*

ii) *$V$ is a pure compactoid;*

iii) *Every strictly decreasing sequence in $\|V\| = \{\|x\| \mid x \in V\}$ converges to zero;*

iv) *There exists an orthogonal sequence $(x_n)_{n\in\mathbf{N}_0}$ in $(V, \|\cdot\|)$ with $\lim_{n\to\infty} x_n = 0$ and $V = \overline{[\{x_n \mid n \in \mathbf{N}_0\}]}$;*

v) *$V$ is $c'$-compact.*

*Proof.* We prove i) $\Rightarrow$ ii) $\Rightarrow$ iii) $\Rightarrow$ iv) $\Rightarrow$ i) and i) $\Rightarrow$ v) $\Rightarrow$ iii).

i) $\Rightarrow$ ii): lemma 5.3.

ii) $\Rightarrow$ iii): Suppose there exists a sequence $(a_n)_{n\in\mathbf{N}_0}$ in $V$ such that $\|a_1\| > \|a_2\| > \ldots > \|a_n\| > \ldots$ and $\lim_{n\to\infty} \|a_n\| = \alpha > 0$. Clearly, $(a_n)_{n\in\mathbf{N}_0}$ is an orthogonal sequence in $(V, \|\cdot\|)$.

Put $U = \{x \in V \mid \|x\| < \alpha\}$. $U$ is open and a.c. Then, for each $F \subset V$ finite, $V \neq U + [F]$. Indeed, suppose there exists $F_0 \subset V$, finite, such that $V = U + [F_0]$. Then for each $n \in \mathbf{N}_0$, $a_n \in U + [F_0]$. Hence, for each $n \in \mathbf{N}_0$, there exist $f_n \in [F_0]$ and $u_n \in U$, such that $a_n = f_n + u_n$. Then for each $n \in \mathbf{N}_0$, $f_n = a_n - u_n$ and $\|f_n\| = \|a_n\|$.

It follows that $(f_n)_{n\in\mathbf{N}_0}$ is an orthogonal sequence in $([F_0], \|\cdot\|)$ but this is a contraction because $[F_0]$ is finite dimensional.

iii) $\Rightarrow$ iv): $A$ is weakly $c'$-compact, hence $A$ is bounded (proposition 2.4). It is no restriction to assume that $\sup_{y\in A} \|y\| \leq 1$.

According to [1], theorem, $(V, \|\cdot\|)$ has an orthogonal base $(s_i)_{i\in I}$. We will prove that $I$ is infinite and countable and that $\lim_{i\in I} \|s_i\| = 0$.

Choose $t \in (0,1)$, and for each $n \in \mathbb{N}_0$, put $I_n = \{i \in I \mid t^n < \|s_i\| \le t^{n-1}\}$.
Then for each $n \in \mathbb{N}_0$, we have $\inf_{i\in I_n} \|s_i\| > t^n$.
Indeed, suppose there exists $n_0 \in \mathbb{N}_0$ such that $\inf_{i\in I_{n_0}} \|s_i\| = t^{n_0}$.
Choose $i_1 \in I_{n_0}$. Then we can find $i_2 \in I_{n_0}$ such that $t^{n_0} < \|s_{i_2}\| < \|s_{i_1}\|$; etc.
Hence, $(\|s_{i_n}\|)_{n\in\mathbb{N}_0}$ is a strictly decreasing sequence, bounded away from zero. This is a contradiction.
Put $N = \{n \in \mathbb{N}_0 \mid \#I_n = \infty\}$. Then $N = \phi$.
Indeed, suppose $N \ne \phi$. Let $n_o \in N$ and put $\alpha = \inf_{i\in I_{n_0}} \|s_i\|$. Clearly, $\alpha > t^{n_0}$.
For each $i \in J = \bigcup_{k=1}^{n_0} I_k$, choose $e_i \in \partial A$, such that $\pi(e_i) = s_i$ and $\|s_i\| \le \|e_i\|$.
For each $i \in I\backslash J$, choose $e_i \in \partial A$, such that $\pi(e_i) = s_i$ and $\|s_i\| \le \|e_i\| < \varepsilon$ with $\varepsilon = \frac{\alpha+t^{n_0}}{2}$.
According to lemma 3.11, $(e_i)_{i\in J}$ is an $(\alpha)$-orthogonal family in $(E, \|\cdot\|)$.
Put $D = \overline{[\{e_i \mid i \in J\}]}$. Then $(e_i)_{i\in J}$ is an $(\alpha)$-orthogonal base of $(D, \|\cdot\|)$ (proposition 3.7.i)).
Define $q : D \to \mathbb{R}_+ : x = \sum_{i\in J} \lambda_i.e_i \ (\lambda_i \to 0) \to \alpha.\max_{i\in J} |\lambda_i|$. Then $q$ is a norm on $D$, equivalent to $\|\cdot\|$ and for each $x \in D$, we have: $\alpha.\|x\| \le q(x) \le \|x\|$. (The inequalities follow from a proof similar to the one in 3.11.)
We can extend $q$ to a norm $\overline{q}$ on $E$, such that for each $x \in E$, $\alpha.\|x\| \le \overline{q}(x) \le \|x\|$ and such that $\overline{q}|_D = q$ ([9], lemma 6.14).
Let $(i_k)_{k\in\mathbb{N}_0}$ be a subset of $I_{n_0}$ and choose $(\eta_k)_{k\in\mathbb{N}_0}$ in $K$ such that $0 < |\eta_1| < |\eta_2| < \ldots < 1$ and $\lim_{n\to\infty} |\eta_k| = 1$.
We define $f : D \to K : \sum_{i\in J} \lambda_i.e_i \to \sum_{k\in\mathbb{N}_0} \lambda_{i_k}.\eta_k$. Clearly, $f$ is linear and well defined.
For $x = \sum_{i\in J} \lambda_i.e_i$ with $\lambda_i \in K$ for each $i \in J$ and with $\lambda_i \to 0$, we have that:

$$|f(\sum_{i\in J} \lambda_i.e_i)| = |\sum_{k\in\mathbb{N}_0} \lambda_{i_k}.\eta_k| \le \max_{k\in\mathbb{N}_0} |\lambda_{i_k}|.|\eta_k| < \max_{j\in J} |\lambda_j|$$
$$\frac{1}{\alpha}\alpha.\max_{j\in J} |\lambda_i| = \frac{1}{\alpha}.q(x) = \frac{1}{\alpha}.\overline{q}(x).$$

Hence, $f$ is continuous.
Choose $\beta > 0$ such that $\frac{1}{\alpha}(1+\beta) < \frac{1}{\varepsilon}$.
We can extend $f \in D'$ to $\overline{f} \in E'$ such that $\overline{f}|_D = f$ and for each $x \in E$:

$$|\overline{f}(x)| \le \frac{1}{\alpha}(1+\beta).\overline{q}(x)$$

(theorem 1.6.).

Then: for each $i \in J \backslash \{i_k \mid k \in \mathbb{N}_0\} : \overline{f}(e_i) = f(e_i) = 0$;

for each $k \in \mathbb{N}_0 : |\overline{f}(e_{i_k})| = |f(e_{i_k})| = |\eta_k| < 1$;

for each $i \in I \backslash J : |\overline{f}(e_i)| \le \frac{1}{\alpha}(1+\beta).\overline{q}(e_i) \le \frac{1}{\alpha}(1+\beta).\|e_i\| < \frac{1}{\varepsilon}.\varepsilon = 1$.

Hence, for each $i \in I : |\overline{f}(e_i)| < 1$, but $\sup\limits_{x\in A} |\overline{f}(x)| = \sup\limits_{i\in I} |\overline{f}(e_i)| = 1$, which is a contradiction.

Hence, for each $n \in \mathbb{N}_0$, $\#I_n < \infty$, so $I$ is countable. As $A = \overline{\text{co}}(e_i \mid i \in I)$ is infinite dimensional, it follows that $I$ is infinite.

Now, $\lim\limits_{i\in I} \|s_i\| = 0$.

Indeed, for each $\delta > 0$, there exists $N_\delta \in \mathbb{N}_0$, such that for each $n > N_\delta$, $t^n < \delta$. Hence, for each $i \in I \backslash \bigcup\limits_{k=1}^{N_\delta} I_k (\ne \phi!)$, $\|s_i\| < \delta$. Hence, as $(s_i)_{i\in I}$ is an orthogonal base of $(V, \|\cdot\|)$, iv) follows.

iv) $\Rightarrow$ i): Let $(x_n)_{n\in\mathbb{N}_0}$ be an orthogonal sequence in $(V, \|\cdot\|)$ with $\lim\limits_{n\to\infty} x_n = 0$ and $V = \overline{[\{x_n \mid n \in \mathbb{N}_0\}]}$. Let $\alpha > 1$.

For each $n \in \mathbb{N}_0$, choose $e_n \in \partial A$, such that $\pi(e_n) = x_n$ and $\|x_n\| \le \|e_n\| < \alpha.\|x_n\|$. According to corollary 4.7, $A = \overline{\text{co}}(e_n \mid n \in \mathbb{N}_0)$ and $\lim\limits_{n\to\infty} e_n = 0$. Hence $A$ is compactoid.

i) $\Rightarrow$ v): Lemma 5.5.

v) $\Rightarrow$ iii): Let $t \in (0,1)$ and suppose there exists a sequence $(a_n)_{n\in\mathbb{N}_0}$ in $V$ such that $\|a_1\| > \|a_2\| > \ldots > \|a_n\| > \ldots$ and $\lim\limits_{n\to\infty} \|a_n\| = \alpha > 0$. Put $D = \overline{[\{a_n | n \in \mathbb{N}_0\}]}$. Clearly, $(a_n)_{n\in\mathbb{N}_0}$ is an orthogonal base of $(D, \|\cdot\|)$. (Proposition 3.7).

(Note that $D = [\{a_n \mid n \in \mathbb{N}_0\}]$, because for each $n \in \mathbb{N}_0$: $\|a_n\| > \alpha$, and the only convergent sums are the finite ones.)

$D$ has a ($t$-)orthogonal complement $F$ (theorem 3.13). Let $(\eta_k)_{k\in\mathbb{N}_0}$ be a sequence in $K$ such that $0 < |\eta_1| < |\eta_2| < \ldots < 1$ and $\lim\limits_{k\to\infty} |\eta_k| = 1$.

Define $q : V \to \mathbb{R}_+ : x = \sum\limits_{n\in\mathbb{N}_0} \lambda_n.a_n + f \; (\lambda_n \in k, f \in F) \to \max\limits_{n\in\mathbb{N}_0} |\lambda_n|.|\eta_n|$.

Clearly, $q$ is a seminorm. Then for $x \in V$,

$$q(x) = q(\sum_{n\in\mathbb{N}_0} \lambda_n.a_n + f) = \max_{n\in\mathbb{N}_0} |\lambda_n|.|\eta_n| < \max_{n\in\mathbb{N}_0} |\lambda_n| \le \frac{1}{\alpha}.\max_{n\in\mathbb{N}_0} \|\lambda_n.a_n\|$$

$$\le \frac{1}{\alpha}.\|\sum_{n\in\mathbb{N}_0} \lambda_n.a_n\| \le \frac{1}{t.\alpha}.t\max\{\|\sum_{n\in\mathbb{N}_0} \lambda_n.a_n\|, \|f\|\} \le \frac{1}{t.\alpha}.\|x\|.$$

Hence, $q$ is continuous.

It is easy to see that for each $x \in V$: $q(x) < 1$, but $\sup_{x\in V} q(x) = \sup_{n\in\mathbb{N}_0} q(a_n) = \sup_{n\in\mathbb{N}_0} |\eta_n| = 1$. But this is a contradiction.

**Remark 5.9.**

1) For results on a.c. weakly $c'$-compact subsets of Banach spaces over a discretely valued field or on finite dimensional a.c. weakly $c'$-compact sets, we refer to [2] and [3].

2) In [3] we also prove that if $K$ is spherically complete and if $A$ is a.c. and weakly $c'$-compact, then $A$ is a compactoid.

3) It was proved in [6] that each a.c. weakly $c'$-compact subset of a Banach space which has a base is a compactoid.

COROLLARY **5.10.** *Suppose $K$ is spherically complete. Let $A$ be closed and weakly $c'$-compact. Then there exists an orthogonal sequence $(e_n)_{n\in\mathbb{N}_0}$ in $\partial A$ such that for each $n \in \mathbb{N}_0$, $\pi(e_n) = s_n$ and $A = \overline{co}(e_n \mid n \in \mathbb{N}_0)$.* (This is another proof of [8], corollary 2.8.)

*Proof.* First we prove that $A$ is edged. Indeed, suppose there exists $x \in A^e \backslash A$. According to [3], proposition 3.5, there exists $f \in E'$ such that $|f(x)| = 1$ and $|f| < 1$ on $A$. As $A$ is weakly compact, $\max_{y\in A} |f(y)| < 1$, and thus, as $A^e = \bigcap_{|\lambda|>1} \lambda A$, $\sup_{y\in A^e} |f(y)| < 1$, a contradiction.

From remark 2 of 5.9, we deduce that that $A$ is a compactoid, hence, $(V, \|\cdot\|)$ has an orthogonal base $(s_n)_{n\in\mathbb{N}_0}$. (Theorem 5.8). For each $n \in \mathbb{N}_0$, there exists $a_n \in \partial A$, such that $\pi(a_n) = s_n$. Put $J_1 = \{n \in \mathbb{N}_0 \mid \|a_n\| = \|s_n\|\}$ and $J_2 = \mathbb{N}_0 \backslash J_1$.

If $n \in J_2$ (i.e. $\|a_n\| > \|s_n\|$), then there exists $y_n \in \overline{A^i}$, such that $\pi(a_n - y_n) = s_n$ and $\|a_n - y_n\| = \|s_n\|$.

Indeed, let $n \in J_2$. For $k \in \mathbb{N}_0$, put $B_{k,n} = \{y \in \overline{A^i} \mid \|a_n - y\| < \|s_n\| + \frac{1}{k}\}$. Obviously, for each $k \in \mathbb{N}_0$, $B_{k,n}$ is convex, closed and non-empty and $B_{k,n+1} \subset B_{k,n}$.

According to [9], theorem 6.15, $\bigcap_{k\in\mathbb{N}_0} B_{k,n} \neq \phi$. Choose $y_n \in \bigcap_{k\in\mathbb{N}_0} B_{k,n}$. Then $\|a_n - y_n\| < \|s_n\| + \frac{1}{k}$, for each $k \in \mathbb{N}_0$, hence, $\|a_n - y_n\| \le \|s_n\|$, and thus $\|a_n - y_n\| = \|s_n\|$.

Clearly, $\pi(a_n - y_n) = s_n$.

Now, for each $n \in J_1$, put $e_n = a_n$;

for each $n \in J_2$, put $e_n = a_n - y_n$.

According to proposition 3.12, the sequence $(e_n)_{n\in \mathbb{N}_0}$ thus constructed is an orthogonal sequence in $((E, \|\cdot\|)$ and according to corollary 4.7, $A = \overline{\text{co}}(e_n \mid n \in \mathbb{N}_0)$.

## REFERENCES

[1] Bayod, J.M. and Martínez-Maurica, J.: *A characterization of the spherically complete normed spaces with a distinguished basis.* Compositio Math. **49** (1983), 143–145.

[2] Borrey, S.: *On the Krein Milman theorem in vector spaces over a non-archimedean valued field $K$.* Indag. Math. (N.S.) **1** (1990), 169–178.

[3] Borrey, S.: *Weakly $c'$-compact subsets of non-archimedean Banach spaces over a spherically complete field $K$.* Collect. Math. **40** (1) (1989), 31–53.

[4] Schikhof, W.H.: *Finite dimensional convexity.* Report 8538, Dept. Math., Catholic University, Nijmegen, The Netherlands (1985).

[5] Schikhof, W.H.: *Topological stability of p-adic compactoids under continuous injections.* Report 8644, Dept. Math., Catholic University, Nijmegen, The Netherlands (1989).

[6] Schikhof, W.H.: *Weak $c'$-compactness in p-adic Banach spaces.* Report 8648, Dept. Math., Catholic University, Nijmegen, The Netherlands (1986).

[7] Schikhof, W.H.: *Locally convex spaces over non-spherically complete valued fields.* Groupe d'étude d'analyse ultramétrique, Paris 1984/85, no. **24**, 33p.

[8] Schikhof, W.H.: *A complementary variant of c-compactness in p-adic functional analysis.* Report 8647, Dept. Math., Catholic University, Nijmegen, The Netherlands (1986).

[9] Van Rooij, A.C.M.: *Notes on p-adic Banach spaces, VI, Convexity.* Report 7725, Dept. Math., Catholic University, Nijmegen, The Netherlands (1977).

[10] Van Rooij, A.C.M.: *Non-archimedean functional analysis.* Marcel Dekker, New York, 1978.

Vrije Universiteit Brussel
Pleinlaan 2
1050 Brussel, Belgium

# $C(E,F)$ AS A DUAL SPACE

N. De Grande-De Kimpe

ABSTRACT. Let $E$ and $F$ be non-archimedean Banach spaces and consider on $L(E,F)$ the topology of uniform convergence on the compactoid subsets of $E$. Then the dual of $L(E,F)$ can be identified with $C(F,E)$, the compact operators from $F$ to $E$.

## 1. INTRODUCTION

Throughout this paper $K$ is a non-archimedean valued field with a non-trivial valuation. We assume that $K$ is complete. By $E$ and $F$ we denote Banach spaces over $K$. Then $L(E,F)$ stands for the vector space of the continuous linear maps from $E$ to $F$. A map $T \in L(E,F)$ is said to be compact if $T(B_E)$, $B_E$ being the unit ball in $E$, is a compactoid subset of $F$. We denote by $C(E,F)$ the vector space of the compact maps from $E$ to $F$.

For all further notions concerning Banach spaces over $K$ we refer to [4].

The aim of this paper is to prove that when $L(E,F)$ is equipped with a suitable topology, its topological dual space can be identified as a vector space with $C(F,E)$. I.e. $(L(E,F))' = C(F,E)$. However, if $E' = \{0\}$ this might not be true. Indeed, taking then $F = K$ we would obtain

$$(L(E,K))' = \{0\} = E = C(K,E)$$

and of course we assume that $E \neq \{0\}$.

In order to avoid complications of this type we shall from now on consider only Banach spaces $E$ and $F$ whose dual spaces $E'$ and $F'$ separate the points of $E$ and $F$. Remark that this is a very mild condition.

## 2. THE TOPOLOGY ON $L(E, F)$

### 2.1. DEFINITIONS

**2.1.1.** We put $c_0(E) = \{(x_n) \subset E;\ \lim_n x_n = 0\}$, and for every $(x_n) \in c_0(E)$ we define a semi-norm $p_{(x_n)}$ on $L(E, F)$ by:

$$p_{(x_n)}(T) = \max_n \|T(x_n)\|, \qquad T \in L(E, F).$$

The family of semi-norms

$$\mathcal{P} = \{p_{(x_n)};\ (x_n) \in c_0(E)\}$$

determines a locally convex topology $\mathcal{T}_{c_0}$ on $L(E, F)$ and we denote by $L_{c_0}(E, F)$ the space $L(E, F)$ with the topology $\mathcal{T}_{c_0}$.

**2.1.2.** For $0 < t < 1$, let $\mathcal{P}'$ be the subfamily of $\mathcal{P}$ defined by:

$$\mathcal{P}' = \{p_{(x_n)};\ (x_n) \in c_0(E) \text{ and } (x_n) \text{ is a } t\text{-orthogonal sequence in } E\}.$$

**2.1.3.** For every compactoid subset $A$ of $E$ we define a semi-norm $p_A$ on $L(E, F)$ by

$$p_A(T) = \sup_{x \in A} \|T(x)\|, \quad T \in L(E, F)$$

and we put

$$\mathcal{P}'' = \{p_A;\ A \subset E, A \text{ is compactoid}\}.$$

Remark that the locally convex topology on $L(E, F)$ determined by the family of semi-norms $\mathcal{P}''$ is called the topology of uniform convergence on the compactoid subsets of $E$.

**2.2. PROPOSITION.** *The families of semi-norms $\mathcal{P}$, $\mathcal{P}'$, and $\mathcal{P}''$ determine the same locally convex topology on $L(E, F)$.*

*Proof.* For $(x_n) \in c_0(E)$ we put $N = \{x_1, x_2, \ldots\}$, $M = \mathrm{Co}(N)$, the absolutely convex hull of $N$, and $\overline{M}$ the closure of $M$ in $E$. Let further $p_M$ and $p_{\overline{M}}$ be defined by

$$p_M(T) = \sup_{x \in M} \|T(x)\|, \quad p_{\overline{M}}(T) = \sup_{x\ in \overline{M}} \|T(x)\|, \qquad T \in L(E, F).$$

Then (see [3] 2.5) $p_{(x_n)} = p_M = p_{\overline{M}}$.
Taking this in mind, the proof is a direct application of the characterization of compactoid sets in Banach spaces (see [4] 4.37).

**2.3.** Remark that in the families $\mathcal{P}$ and $\mathcal{P}'$ it suffices to consider those sequences $(x_n)$ for which $x_n \neq 0$, $n = 1, 2, 3, \ldots$.
Omitting sequences with zero-terms does indeed not alter the topology.

## 3. THE DUAL OF $L_{c_0}(E, F)$

**3.1.** LEMMA. *Consider on $c_0(F)$ the norm $\|(y_n)\| = \max_n \|y_n\|$, $(y_n) \in c_0(F)$. Then the topological dual space of the Banach space $c_0(F)$ is given by*

$$(c_0(F))' = \ell^\infty(F') = \{(g_n) \subset F';\ \sup_n \|g_n\| < \infty\}.$$

*Explicitely: $f \in (c_0(F))'$ if and only if there exists a $(g_n) \in \ell^\infty(F')$ such that the value of $f$ at $(y_n) \in c_0(F)$ is $f((y_n)) = \sum_{n=1}^{\infty} g_n(y_n)$.*
*Proof.* [4] 3.Q.

**3.2.** PROPOSITION. *Suppose $K$ is spherically complete. Then $\phi \in (L_{c_0}(E, F))'$ if and only if there exist $(x_n) \in c_0(E)$ and $(g_n) \in \ell^\infty(F')$ such that for all $T \in L(E, F)$ we have*

$$(*) \qquad \phi(T) = \sum_{n=1}^{\infty} g_n(T(x_n)).$$

*Proof.* i) If $\phi$ is defined by $(*)$ then it is obviously linear. It is also continuous because

$$|\phi(T)| \leq \max_n |g_n(T(x_n))| \leq \sup_n \|g_n\| p_{(x_n)}(T).$$

ii) Take $\phi \in (L_{c_0}(E, F))'$. Then there exists $(x_n) \in c_0(E)$ such that

$$(**) \qquad |\phi(T)| \leq p_{(x_n)}(T) = \max_n \|T(x_n)\|, \text{ for all } T \in L(E, F).$$

We now consider the subspace $X$ of $c_0(F)$ defined by

$$X = \{(T(x_n));\ T \in L(E, F)\},$$

and equipped with the norm induced by $c_0(F)$.
If $S, T \in L(E, F)$ and $(S(x_n)) = (T(x_n))$, then it follows from (**) that $\phi(S) = \phi(T)$.
Therefore the map

$$\eta : X \to K : (T(x_n)) \to \phi(T)$$

is well defined. It is linear and, again by (**), it is continuous.
Since $K$ is spherically complete the map $\eta$ has a linear continuous extension

$$\overline{\eta} : c_0(F) \to K.$$

By lemma 3.1 there exists an element $(g_n) \in \ell^\infty(F')$ such that for all $(y_n) \in c_0(F)$ we have

$$\overline{\eta}((y_n)) = \sum_{n=1}^{\infty} g_n(y_n).$$

This gives

$$\phi(T) = \eta((T(x_n)) = \overline{\eta}((T(x_n)) = \sum_{n=1}^{\infty} g_n(T(x_n))$$

and we are done.

## 4. $C(F, E)$ AS A DUAL SPACE

**4.1.** DEFINITION. Let $\mathcal{F}(F, F)$ stand for the subspace of $L(F, F)$ consisting of the finite rank operators. For every $T \in \mathcal{F}(F, F)$ there exist $k \in \mathbb{N}$, $f_1, f_2, \ldots, f_k \in F'$, and $a_1, a_2, \ldots, a_k \in F$ such that for all $x \in F$

$$(*) \qquad T(x) = \sum_{n=1}^{k} f_n(x) a_n.$$

Then we define, as in the classical theory, the trace of $T$ by

$$\operatorname{tr}(T) = \sum_{n=1}^{k} f_n(a_n).$$

As $\operatorname{tr}(T)$ is independent of the representation of $T$ as $(*)$, we have a map

$$\operatorname{tr} : \mathcal{F}(F, F) \to K : T \to \operatorname{tr}(T)$$

which clearly is linear.

**4.2. LEMMA.** *If on $\mathcal{F}(F, F)$ we consider the topology induced by the norm topology on $L(F, F)$, then the map tr is continuous.*

*Proof.* Take $T \in \mathcal{F}(F, F)$ and let $e_1, e_2, \ldots, e_k$ be a $t$-orthogonal basis for $\mathrm{Im}(T)$, $0 < t < 1$. Then there exist $f_1, f_2, \ldots, f_k \in F'$ such that

$$T(x) = \sum_{n=1}^{k} f_n(x)e_n, \text{ for all } x \in F.$$

Now

$$|\mathrm{tr}(T)| = |\sum_{n=1}^{k} f_n(e_n)| \leq \max_n \|f_n\| \cdot \|e_n\|$$

$$\max_n (( \sup_{\|x\| \leq 1} |f_n(x)|) \cdot \|e_n\|)$$

$$= \sup_{\|x\| \leq 1} (\max_n |f_n(x)| \cdot \|e_n\|) \leq 1/t \cdot \sup_{\|x\| \leq 1} \|T(x)\|.$$

Hence for all $T \in \mathcal{F}(F, F)$, $|\mathrm{tr}(T)| \leq 1/t \cdot \|T\|$ which proves the continuity of tr.

**4.3. TRACE OF A COMPACT OPERATOR**

If on $L(F, F)$ we consider the norm topology we know that $C(F, F)$ is the closure of $\mathcal{F}(F, F)$ in $L(F, F)$. (see [4] 4.40). Hence, by lemma 4.2, the map tr has a unique extension to $C(F, F)$ which we still denote by tr. So every $T \in C(F, F)$ has a trace $\mathrm{tr}(T)$.

**4.4. EXPLICIT FORM OF $\mathrm{tr}(T)$ FOR $T \in C(F, F)$**

From [4] 4.40 it follows easily that for every $T \in C(F, F)$ there exist sequences $(a_n) \in c_0(F)$ and $(f_n) \in \ell^\infty(F')$ such that

$$(**) \qquad T(x) = \sum_{n=1}^{\infty} f_n(x)a_n, \qquad \text{for all } x \in F.$$

For $k = 1, 2, \ldots$ we put

$$T_k(x) = \sum_{n=1}^{k} f_n(x)a_n, \qquad x \in F.$$

Then $T_k \in \mathcal{F}(F, F)$ for all $k$, and $\lim_k T_k = T$ in $L(F, F)$ with the norm topology.

Hence by lemma 4.2, $\operatorname{tr}(T) = \lim_k \operatorname{tr}(T_k)$ which gives (see 4.1):

$$\operatorname{tr}(T) = \sum_{n=1}^{\infty} f_n(a_n).$$

Moreover, it follows that this value of $\operatorname{tr}(T)$ does not depend on the representation of $T$ as $(**)$.
In particular we have

**4.5.** If we fix a $T \in C(F, E)$, then for every $S \in L(E, F)$ we have $S \circ T \in C(F, F)$ and so $\operatorname{tr}(S \circ T)$ exists.
Write $T$ as

$$T(y) = \sum_{n=1}^{\infty} f_n(y) a_n, \qquad y \in F$$

with $(f_n) \in \ell^\infty(E')$ and $(a_n) \in c_0(E)$. ([4] 4.40).
Then, for $y \in F$

$$S \circ T(y) = \sum_{n=1}^{\infty} f_n(y) S(a_n)$$

and

$$\operatorname{tr}(S \circ T) = \sum_{n=1}^{\infty} f_n(S(a_n)),$$

where the value of $\operatorname{tr}(S \circ T)$ does not depend on the representation of $T$.

Now we are ready to prove the main result.

**4.6. THEOREM.** *If $K$ is spherically complete, then the topological dual space of $L_{c_0}(E, F)$ can be identified, as a vector space, with $C(F, E)$.*

*Proof.* We need to construct a linear bijection

$$U : C(F, E) \to (L_{c_0}(E, F))'.$$

*Definition of $U$:*
For $T \in C(F, E)$ we define (see 4.5)

$$U_T : L_{c_0}(E, F) \to K : S \to \operatorname{tr}(S \circ T).$$

Obviously $U_T$ is a linear map. Also $U_T$ is continuous because, with the notations of 4.5, we have

$$\begin{aligned}|U_T(S)| = |\mathrm{tr}(S \circ T)| &= |\sum_{n=1}^{\infty} f_n(S(a_n))| \\ &\leq \sup_n \|f_n\| \cdot \max_n \|S(a_n)\| \\ &= \sup_n \|f_n\| \cdot p_{(a_n)}(S).\end{aligned}$$

Now define

$$U : C(F, E) \to (L_{c_0}(E, F))'$$

by $U(T) = U_T$.
Then $U$ is a linear map.

*U is injective:*
Indeed, if $U(T) = 0$ we have, still with the notations of 4.5,

$$\sum_{n=1}^{\infty} f_n(S(a_n)) = 0, \quad \text{for all } S \in L(E, F).$$

In particular this holds for every $S \in L(E, F)$ of the form

$$S(x) = g(x) \cdot b, \quad \text{with } g \in E' \text{ and } b \in F.$$

Hence

$$\sum_{n=1}^{\infty} f_n(b) g(a_n) = 0, \quad \text{for all } g \in E', b \in F.$$

Now

$$\sum_{n=1}^{\infty} f_n(b) g(a_n) = g(\sum_{n=1}^{\infty} f_n(b) a_n) = g(T(b)).$$

Thus $g(T(b)) = 0$ for all $g \in E'$ and all $b \in F$.
It follows that $T = 0$.

*U is surjective:*
Take $\phi \in (L_{c_0}(E, F))'$. Then, by proposition 3.2 there exist $(x_n) \in c_0(E)$ and $(g_n) \in \ell^\infty(F')$ such that for all $S \in L(E, F)$

$$\phi(S) = \sum_{n=1}^{\infty} g_n(S(x_n)).$$

Putting

$$T(y) = \sum_{n=1}^{\infty} g_n(y)x_n$$

we have that $T \in C(F, E)$, and for all $S \in L(E, F)$

$$U_T(S) = \text{tr}(S \circ T) = \sum_{n=1}^{\infty} g_n(S(x_n)) = \phi(S).$$

So $U(T) = \phi$.

## 5. THE NON-SPHERICALLY COMPLETE CASE

**5.1. Remark.** An inspection of the proof of theorem 3.2 shows that the condition "$K$ is spherically complete" is not used to its full extent.

Let us recall the situation.

Fix an $(x_n) \in c_0(E)$ and remark first that we can limit ourselves to those $(x_n)$ that are $t$-orthogonal and with $x_n \neq 0$ for all $n$ (see proposition 2.2 and remark 2.3).

Then consider the vector space

$$X = \{(T(x_n));\ T \in L(E, F)\} \subset c_0(F)$$

where $X$ is equipped with the topology induced by the norm on $c_0(F)$.

Further we had a continuous linear map $\eta : X \to K$ and we used the spherical completeness of $K$ in order to obtain a continuous linear extension of $\eta$ to $\overline{\eta} : c_0(F) \to K$.

Once this extension is obtained the proof of proposition 3.2 goes through and then also the proof of the main result, theorem 4.6.

Suppose now that $K$ is not spherically complete.

Then in some cases the extension $\overline{\eta}$ still exists. We'll show here that this is the case when $E$ has a basis or when $F$ is of countable type.

This leads us to

**5.2. THEOREM.** *Suppose $K$ is not spherically complete. Then, if $E$ has a basis or $F$ is of countable type, we have*

$$(L_{c_0}(E, F))' = C(F, E)$$

*as vector spaces.*

*Proof.* i) If $F$ is of countable type, then it is easy to see that $c_0(F)$ is of countable type as well. In that case (see [4] 3.16 vi)), for every subspace $X$ of $c_0(F)$ and every $\eta \in X'$ there exists an extension $\overline{\eta} \in (c_0(F))'$
If $E$ has a basis, then the existence of the extension follows from the fact that in this case $X$ is dense in $c_0(F)$. This will be proved in 5.4.

**5.3. LEMMA.** *Let $X$ be a subspace of the Banach space $c_0(F)$ which has the following properties*

i) *For all $y \in F$ and all $k \in \mathbf{N}$ there exists a $(z_n) \in X$ with $z_k = y$.*
ii) *If $(z_n) \in X$, then the set of its sections*

$$\{(z_1, z_2, \ldots, z_k, 0, 0, \ldots);\ k = 1, 2, \ldots\}$$

*is a subset of $X$.*

*Then $X$ is dense in $c_0(F)$.*

*Proof.* Take $(y_n) \in c_0(F)$.
By i) there exists a $(z_n^1) \in X$ of the form

$$(z_n^1) = (y_1, z_2^1, z_3^1, \ldots).$$

Then by ii) we have that

$$(y_1, 0, 0, \ldots) \in X.$$

Again by i), there exists a $(z_n^2) \in X$ of the form

$$(z_n^2) = (z_1^2, y_2, z_3^2, z_4^2, \ldots).$$

Then by ii) we have that

$$(z_1^2, 0, 0, \ldots) \in X \quad \text{and} \quad (z_1^2, y_2, 0, 0, \ldots) \in X.$$

Since $X$ is a vector space this gives

$$(0, y_2, 0, 0, \ldots) \in X,$$

and then also

$$(y_1, y_2, 0, 0, \ldots) \in X.$$

Continuing this type of argument we arrive at

$$(y_1, y_2, y_3, \ldots, y_k, 0, 0, 0, \ldots) \in X, \text{ for } k = 1, 2, 3, \ldots$$

On the other hand it is clear that in $c_0(F)$

$$(y_n) = \lim_k (y_1, y_2, \ldots, y_k, 0, 0, \ldots).$$

Hence $X$ is dense in $c_0(F)$.

**5.4. LEMMA.** *Let $(x_n) \in c_0(E)$ be such that $(x_n)$ is t-orthogonal* $(0 < t < 1)$ *and $x_n \neq 0$ for $n = 1, 2, 3, \ldots$.*
*Put $X = \{(T(x_n));\ T \in L(E, F)\} \subset c_0(F)$. Then*
a) *$X$ satisfies condition* i) *in lemma* 5.3.
b) *If $E$ has a basis then $X$ also satisfies condition* ii) *in* 5.3 *and therefore $X$ is dense in $c_0(F)$.*

*Proof.* a) Take $y \in F$ and $k \in \mathbb{N}$. Then there exists $f \in E'$ such that $f(x_k) = 1$. We now define $T \in L(E, F)$ by $T(x) = f(x) \cdot y$, $x \in E$. Then $T(x_k) = y$ as required.
b) In order to satisfy condition ii) we need to find for every $T \in L(E, F)$ and every $k \in N$ a map $S \in L(E, F)$ such that

$$(*) \qquad \begin{aligned} &S(x_n) = T(x_n), \quad \text{for } n = 1, 2, \ldots, k \quad \text{and} \\ &S(x_n) = 0, \quad \text{for } n > k. \end{aligned}$$

We denote by $D$ the closure in $E$ of the linear hull of the set $\{x_1, x_2, \ldots\}$.
I.e. $D = \overline{[x_1, x_2, \ldots]}$.
Then $D$ is of countable type and $x_1, x_2, \ldots$ is a $t$-orthogonal basis for $D$.
Let $P_k$ stand for the continuous projection

$$P_k : D \to D_k = [x_1, x_2, \ldots, x_k] : \sum_{n=1}^{\infty} \lambda_n x_n \to \sum_{n=1}^{k} \lambda_n x_n.$$

Finally we write $T_k$ for the restriction of $T$ to $D_k$.
Since $E$ has a basis the subspace $D$ is complemented in $E$ (see [4] 3.18). We denote by $P$ the continuous projection from $E$ onto $D$.
Now it is clear that the continuous linear map $S = T_k \circ P_k \circ P$ satisfies the required condition $(*)$.

**5.5. PROBLEM.** Find the weakest conditions on $E$ and (or) $F$ under which the statement in theorem 5.2 holds.

## 6. SOME SPECIAL CASES

**6.1.** If $F = K$ then the theorem holds without extra conditions on $K$. Since in that case

$$(L_{c_0}(E,F))' = E'_c = E'$$

with the topology $\mathcal{T}_c$ of uniform convergence on compactoid subsets of $E$, and

$$C(K,E) = E,$$

the theorem means that the topology $\mathcal{T}_c$ is compatible with the dual pair $(E', E)$.

**6.2.** It is interesting to apply the theorem to situations in which all the continuous linear maps are compact.
First remark that this is the case whenever $E$ or $F$ is finite dimensional. Hence in this case we have

$$(L_{c_0}(E,F))' = L(F,E),$$

without extra conditions on $K$.
Further recall that (see [2] theorem p.246) if $E$ is weakly projective or $F$ is weakly injective and $L(E,F) = C(E,F)$ then $E$ or $F$ is finite dimensional. This is for instance the case when the valuation on $K$ is discrete ([4] 4.14 and 5.16).
Therefore we now can restrict ourselves to the situation in which the valuation on $K$ is dense.

Take now $E = \ell^\infty$ and $F = c_0$. Then the theorem holds without further condition on $K$ ($F$ is of countable type), and since $L(\ell^\infty, c_0) = C(\ell^\infty, c_0)$ (see [4] 5.19) we get

$$(C_{c_0}(\ell^\infty, c_0))' = C(c_0, \ell^\infty).$$

If $E = c_0$ and $F = \ell^\infty$ again the theorem holds without further condition on $K$ ($E$ has a basis) and we get

$$(L_{c_0}(c_0, \ell^\infty))' = L(\ell^\infty, c_0).$$

## 7. THE TOPOLOGICAL IDENTIFICATION

In this section we investigate under which conditions the linear isomorphism (from 4.6 and 5.2)

$$U : C(F,E) \to (L_{c_0}(E,F))'$$

is a (topological) homeomorphism for the "natural topologies" on both spaces.
In order to find out what these " natural topologies" are and what the conditions might be we first return to the special case $F = K$.

### 7.1. THE CASE $F = K$

Here, with the notations of 6.1, we have

$$U : E \to (E'_c)'.$$

Since the canonical norm-topology on $E'$ is finer than $\mathcal{T}_c$ the space $(E'_c)'$ can be identified with a subspace of $E''$. Hence the natural topology on $(E'_c)'$ is the topology induced by the canonical norm on $E''$.
Let us now look more closely to the map $U$.
We have that $C(K,E)$ and $E$ are identified as Banach spaces by

$$C(K,E) \to E : T \to T(1) = t.$$

Then, for $f \in E' = L(E,K) : \mathrm{tr}(f \circ T) = f(t)$.
Hence the map $U$ is defined by

$$U(t) = U_t, \quad \text{where } U_t(f) = f(t).$$

This means that

$$U(E) = (E'_c)' = J_E(E)$$

where $J_E$ is the canonical map from $E$ into $E''$.
From [4] p.60 we then obtain that $U$ is continuous and that "$U$ is open" means exactly that $E$ is topologically pseudoreflexive.

**7.2.** Remark that when $K$ is spherically complete every Banach space over $K$ is topologically pseudoreflexive. This follows easily from [5] 4.14 and 4.15.

### 7.3. THE GENERAL CASE

The natural topology on $C(F,E)$ is the topology induced by the canonical norm on $L(F,E)$, recall that $C(F,E) \subset L(F,E)$.
On the other hand, since the topology $\mathcal{T}_{c_0}$ on $L(E,F)$ is weaker than the norm topology $\mathcal{T}$ we can consider $(L_{c_0}(E,F))'$ as a subspace of $(L_{\mathcal{T}}(E,F))'$. So the natural topology on $(L_{c_0}(E,F))'$ is the topology induced by the canonical norm on $(L_{\mathcal{T}}(E,F))'$.
Concretely this means that we take:

for $T \in C(F,E)$, $\|T\| = \sup\{\|T(y)\|;\ y \in F,\ \|y\| \leq 1\}$

and

for $\phi \in (L_{c_0}(E,F))'$, $\|\phi\| = \sup \||\phi(S)|;\ S \in L(E,F),\ \|S\| \leq 1\}$.

We then have:

**7.4. THEOREM.** *Suppose $K$ is spherically complete. Then the normed spaces $(L_{c_0}(E,F))'$ and $C(F,E)$ are linearly homeomorphic.*
*This implies in particular that $(L_{c_0}(E,F))'$ is a Banach space.*

*Proof.* We need to prove that the map $U$ from theorem 4.6 is continuous and open.

*$U$ is continuous:*
First remark that it is not hard to see that every $T \in C(F,E)$ can be written as $T(y) = \sum\limits_{n=1}^{\infty} f_n(y)a_n$, $y \in F$; with $(f_n) \in \ell^\infty(F')$ and $(a_n) \in c_0(E)$ where the sequence $(a_n)$ is $t$-orthogonal $(0 < t < 1)$. Then, for such $T$ and for $S \in L(E,F)$ with $\|S\| \leq 1$ we have as before

$$\begin{aligned}|\mathrm{tr}(S \circ T)| &= |\sum_{n=1}^{\infty} f_n(S(a_n))| \leq \max_n |f_n(S(a_n))| \\ &\leq \max_n \|f_n\|\ \|a_n\| = \max_n(\sup_{\|y\|\leq 1} |f_n(y)|\ \|a_n\|) \\ &= \sup_{\|y\|\leq 1} (\max_n |f_n(y)|\ \|a_n\|).\end{aligned}$$

On the other hand

$$\|T\| = \sup\{\|\sum_{n=1}^{\infty} f_n(y)a_n\|;\ \|y\| \leq 1\} \geq t \sup\{\max_n |f_n(y)| \cdot \|a_n\|;\ \|y\| \leq 1\}.$$

Hence

$$|\mathrm{tr}(S \circ T)| \leq 1/t\|T\|.$$

This gives

$$\|U(T)\| = \sup\{|U_T(S)| = |\mathrm{tr}(S \circ T)|;\ \|S\| \leq 1\} \leq 1/t\|T\|.$$

*U is open:*

For $f \in E'$, with $\|f\| \leq 1$ and $y \in F$ with $\|y\| \leq 1$, we define $S \in L(E, F)$ by $S(x) = f(x)y$, $x \in E$.

Then $\|S\| \leq 1$ and for $T \in C(F, E)$ written as above, we have

$$\mathrm{tr}(S \circ T) = \sum_{n=1}^{\infty} f(a_n) f_n(y).$$

This gives

$$\|U(T)\| \geq \sup\{|\sum_{n=1}^{\infty} f(a_n) f_n(y)|;\ \|f\| \leq 1,\ \|y\| \leq 1\}.$$

On the other hand, since $E$ is topologically pseudoreflexive, there exists $C > 0$ such that for all $x \in E$

$$\|x\| \leq C\|J_E(x)\| = C \sup\{|f(x)|;\ f \in E',\ \|f\| \leq 1\}.$$

This gives

$$\begin{aligned}\|T\| &= \sup\{\|T(y)\|;\ y \in F, \|y\| \leq 1\} \\ &\leq C \sup\{|\sum_{n=1}^{\infty} f_n(y) f(a_n)|;\ \|y\| \leq 1,\ \|f\| \leq 1\}\end{aligned}$$

and we conclude that

$$\|U(T)\| \geq 1/C \cdot \|T\|.$$

With the same proof we get from 5.2

**7.5. THEOREM.** *Suppose $K$ is not spherically complete and $E$ is topologically pseudoreflexive. Then, if $E$ has a basis or $F$ is of countable type, the normed spaces $(L_{c_0}(E, F))'$ and $C(F, E)$ are linearly homeomorphic.*

## 8. SOME IDEAS FOR FURTHER STUDY

**8.1.** If $E$ and $F$ are locally convex spaces there is no problem in defining the topology $\mathcal{T}_{c_0}$ on $L(E,F)$. Of course it will not be the topology of uniform convergence on compactoid sets unless $E$ is a metric space.
For the description of $(L_{c_0}(E,F))'$ as a space of operators from $F$ to $E$ one probably will have to look at the types of operators introduced in [1].

**8.2.** An other way to generalize the theorems in this paper is to replace $c_0$ by a -more or less- arbitrary sequence space $\Lambda$, and study the operators obtained in the dual of $L(E,F)$ with the topology $\mathcal{T}_\Lambda$.

Each of these ideas will be developed later on.

## REFERENCES

[1] De Grande-De Kimpe, N.: *On spaces of operators between locally K-convex spaces.* Nederl. Akad. Wetensch. Proc. Ser. A **75** = Indag. Math. **34** (1972), 113–129.

[2] De Grande-De Kimpe, N.: *Non-archimedean Banach spaces for which all the operators are compact.* Nieuw Arch. Wisk. **22** (1974), 241–248.

[3] De Grande-De Kimpe, N.: *Non-archimedean topologies of countable type and associated operators.* Nederl. Akad. Wetensch. Indag. Math. **49** (1987), 15–28.

[4] Van Rooij, A.C.M.: *Non-archimedean functional analysis.* Marcel Dekker, New York, 1978.

[5] Van Tiel, J.: *Espaces localement K-convexes III.* Nederl. Akad. Wetensch. Proc. Ser. A **68** = Indag. Math. **27** (1965), 273–289 (1965).

DEPARTMENT OF MATHEMATICS
VRIJE UNIVERSITEIT BRUSSEL
PLEINLAAN 2 (10F7), 1050 BRUSSEL, BELGIUM

# NON INTEGRALLY CLOSED ALGEBRAS $H(D)$

Alain Escassut and Bertin Diarra

**ABSTRACT.** Let $(K, |.|)$ be an algebraically closed complete ultrametric field. Let $D$ be a closed bounded infraconnected set in $K$, and let $H(D)$ be the Banach algebra of the analytic elements in $D$. Let $f$ be a power series whose radius of convergence is $r$. If for some $s \in \mathbb{N}^*$, $f^s$ has a radius of convergence strictly greater than $r$, and if $f^s$ has less than $s$ zeros in $C(0,r)$ then $f$ does not belong to $H(d(0,r^-))$.

Let $a$ be an integer, prime to the residue characteristic $p$ of $K$ if $p \neq 0$. The function $\sqrt[a]{1+x}$ does not belong to $H(d(0,1^-))$.

Assume $D$ to have a $T$-filter $F$ with a non empty beach or a $T$-filter with center and with diameter in $|K|$. Then $H(D)$ is not integrally closed.

The article [E$_6$] by A. Escassut (to appear in Communications in Algebra) concerns the case of the integrally closed algebras $H(D)$.

## §1. INTRODUCTION AND THEOREMS

Let $K$ be an algebraically closed complete ultrametric field whose absolute value is denoted by $| \,.\, |$.

Let $D$ be a subset of $K$ whose closure is denoted by $\overline{D}$. We denote by $R(D)$ the $K$-algebra of the rational functions $h(x) \in K(x)$ with no pole in $D$, and we denote by $H(D)$ the completion of $R(D)$ for the topology of uniform convergence on $D$. The elements of $H(D)$ are called the Analytic Elements on $D$[$K_1$, $K_2$].

Let us recall that the topologies of $R(D)$ and $H(D)$ are determined by the uniform convergence norm, denoted by $\| \,.\, \|_D$, if and only if $D$ is closed and bounded [E$_1$,E$_2$].

More generally $H(D)$ is a $K$-subalgebra of $K^D$ if and only if $D$ satisfies the two following conditions [E$_1$,E$_2$].

a) either $D$ or the interior of $K\backslash D$ is bounded,
b) $\overline{D}\backslash D$ is included in the interior of $\overline{D}$.

Finally let us recall that an algebra $H(D)$ has no idempotent different from 0 and 1 if and only if $D$ is infraconnected [E1,E2]. A set $D$ is said to be *infraconnected* if for every $a \in D$ the mapping from $D$ to $\mathbb{R}_+$ $x \to |x-a|$ has an image whose closure is an interval.

The definition of the $T$-filters [$E_4$] involved in Theorem 4 will be recalled below.

Given an algebra $H(D)$ with no idempotent different from 0 and 1, the three following conditions are equivalent

i) every ideal is generated by a polynomial whose zeros belong to $D$,
ii) $H(D)$ is noetherian,
iii) $D$ is open with no $T$-filter.

In [E6] we showed that if $D$ just has no $T$-filter then $H(D)$ is integrally closed (even when $D$ is not open).

Here we will show that *most of the $T$-filters do provide non integrally closed algebras $H(D)$.*

We will first obtain results on the convergence radius of the $n$-th power of a power series $f$, in connection with the number of zeros of $f$, and we will show that the function $\sqrt[p]{1+x}$ never is an analytiuc element in the disk $|x| < 1$ whatever the characteristic of $K$. These results are obtained thanks to Motzkin's factorization into singular factors [M] which have been much helpful. (The idea of using this factorization is linked to M.C. Sarmant's work on the Motzkin singular factors seen as Meromorphic Products [S1].)

NOTATIONS: Let $a \in K$ and let $r \in \mathbb{R}_+$.

$d(a,r)$ denotes the disk $\{x \in K \mid |x-a| \leq r\}$
$d(a,r^-)$ denotes the disk $\{x \in K \mid |x-a| < r\}$
and $C(a,r)$ denotes the circle $\{x \in K \mid |x-a| = r\}$;

Every disk $d(b,r^-)$ with $b \in C(a,r)$ is called *a class of* $C(a,r)$.

Let $D$ be an infraconnected set such that $H(D)$ is a $K$-algebra and let $f \in H(D)$. Then $f$ is said to be *semi-invertible* if it factorizes in the form $Pg$ with $P$ a polynomial whose zeros belong to $D$ and $g$ an element invertible in $H(D)$.

Let $a \in D$ and let $R$ be the diameter of $D(R = +\infty$ if $D$ is not bounded). $\widetilde{D}$ denotes the disk $d(a, R)$ if $R < +\infty$ and $K$ if $R = +\infty$. Then $\widetilde{D} \backslash \overline{D}$ admits a unique partition in the form $(T_i)_{i \in I}$ such that for each $i \in I$, $T_i$ is a disk $d(a_i, r_i^-)$ where $r_i$ is the distance from $a_i$ to $D$. The $T_i$ are called *the holes* of $D$. (For example the holes of $d(0, 1^-)$ are the disks $d(a, 1^-)$ with $|a| = 1$.)

We are now able to state our Theorems.

**THEOREM 1.** *Let $D$ be a closed bounded infraconnected set and let $f \in H(D)$ be semi-invertible. Let $T$ be a hole of $D$. We assume $f^s$ to have a continuation to an element of $H(D \cup T)$ for some $s \in \mathbf{N}^*$, and we assume $f$ not to have a continuation to an element of $H(D \cup T)$. Then the number of zeros of $f^s$ in $T$ is different from zero and is a multiple of $s$ (taking multiplicities into account).*

The following Theorem 2 is a direct consequence of Theorem 1.

**THEOREM 2.** *Let $f \in H(d(0, r^-)) \backslash H(d(0, r))$ be such that for some $s \in \mathbf{N}^*$, $f^s \in H(d(0, r))$. Then $f$ has continuation to an analytic element in a set $D$ in the form $d(0, r) \backslash \left( \bigcup_{i=1}^{t} d(a_i, r^-) \right)$ with $|a_i| = r = |a_i - a_j|$ whenever $i \neq j$, such that for each $i = 1, \ldots, t$ the number of zeros of $f^s$ in $d(a_i, r_i)$ is different from zero and multiple of $s$.*

**COROLLARY 1.** *Let $f$ be a power series whose radius of convergence is $r$. If for some $s \in \mathbf{N}^*$, $f^s$ has a radius of convergence greater than $r$ and if $f^s$ has less than $s$ zeros in $C(0, r)$ then $f$ does not belong to $H(d(0, r^-))$.*

We will denote by $\mathbf{A}(d(0, 1^-))$ the algebra of the power series convergent for $|x| < 1$.

We now have got to recall the definition of $\sqrt[q]{u}$ when $u \in d(1, 1^-)$.

*In all the following, $q$ will denote an integer $\geq 2$, prime to the residue characteristic $p$ if $p \neq 0$.*

When considering the coefficients of $(1 + x)^q = 1 + qx + \sum_{j=2}^{q} \lambda_j x^j$ we see that $|q| = 1$ while $|\lambda_j| \leq 1$ whenever $j = 2, \ldots, q$ hence it is easily seen that the mapping $x \rightarrow (1+x)^q$ defines a bijection from $d(0, 1^-)$ onto $d(1, 1^-)$, hence the mapping $u \rightarrow u^q$

defines a bijection from $d(1,1^-)$ onto $d(1,1^-)$. We denote by $\sqrt[q]{\ }$ the reciprocal mapping from $d(1,1^-)$ onto $d(1,1^-)$ and we put $\psi_q(x) = \sqrt[q]{1+x}$ for $x \in d(0,1^-)$.

**THEOREM 3.** *$\psi_q$ belongs to $\mathbf{A}(d(0,1^-))$ but does not belong to $H(d(0,1^-))$.*

**Remark 1.** $(\sqrt[q]{u})^q = \sqrt[q]{u^q} = u$ whenever $u \in d(1,1^-)$.

**Remark 2.** Since $\psi_q \in \mathbf{A}(d(0,1^-))$, obviously $\psi_q \in H(d(0,r))$ for every $r \in ]0,1[$. Now, if $D$ is a closed bounded set in $K$ and if $h \in H(D)$ is such that $\|h\|_D < 1$, then $\psi_q \circ h \in H(D)$.

In other words if $g \in H(D)$ and if $\|g-1\|_D < 1$ then $\sqrt[q]{g}$ belongs to $H(D)$.

**Remark 3.** Theorems 1 and 2 couldn't be significantly improved as shows the following example. Let $q$ be an integer prime to $p$, let $a, b \in d(1,1^-)$ with $a \neq b$, and let $P(x) = (a-x)^{q-1}(b-x)$. It is easily seen that $|P(x)-1| < 1$ whenever $x \in d(0,1^-)$ hence can we consider $f(x) = \sqrt[q]{P(x)}$. We will first show that $f \in H(d(0,1^-)) \setminus H(d(0,1))$. Indeed by Remark 1, $f(x) = \sqrt[q]{(a-x)^q\left(\frac{b-x}{a-x}\right)} = (a-x)\sqrt[q]{1+\frac{b-a}{a-x}} =$

$$(a-x)\sum_{n=0}^{\infty}\binom{\frac{1}{q}}{n}\left(\frac{b-a}{a-x}\right)^n = a-x+\frac{1}{q}(b-a)+\sum_{j=1}^{\infty}\binom{\frac{1}{q}}{j+1}\frac{(b-a)^{j+1}}{(a-x)^j}.$$

This is just a Mittag-Leffler series $f_0 + f_1 \in H(d(0,1^-))$ with

$$f_0 = -x+a+\frac{1}{q}(b-a) \in H(d(0,1)) \text{ and } f_1 = -\sum_{j=1}^{\infty}\binom{\frac{1}{q}}{j+1}\frac{(a-b)^{j+1}}{(x-a)^j} \in H_0(K\setminus d(1,1^-)).$$

Thus we see that $f \in H(d(0,1^-))$ and more precisely $f \in H(K\setminus d(1,1^-))$ but $f \notin H(d(0,1))$, while $f^q$ has exactly $q$ zeros in $d(1,1^-)$.

**Remark 4.** In [S$_2$] Marie-Claude Sarmant and Alain Escassut gave another proof of the fact that $\psi_q \notin H(d(0,1^-))$ in the case when $K$ has characteristic zero by considering the differential equation $y' = \frac{y}{q(1+x)}$.

As a partial reciprocal of Theorem 1 in [E$_6$] the following Theorem 4 shows that many $T$-filters make sets $D$ such that $H(D)$ is not integrally closed.

We denote by $|K|$ the set of the $|x|$, $x \in K$.

**THEOREM 4.** *Let $D$ be an infraconnected set such that $H(D)$ is a $K$-subalgebra of $K^D$. We assume $D$ to have a $T$-filter $F$ with center which satisfies at least one of the two following conditions $\alpha$) and $\beta$).*

*$\alpha$) The beach of $F$ is not empty.*

*$\beta$) The diameter of $F$ belongs to $|K|$.*

*Then $H(D)$ is not integrally closed.*

**COROLLARY 2.** *We assume $K$ to be spherically complete and provided with a valuation group equal to $\mathbb{R}$. Then an algebra $H(D)$ is integrally closed if and only if the set $D$ is infraconnected with no $T$-filter.*

The conclusion obtained in Corollary 2 is clearly the one we have been looking for in the general case when $K$ is not supposed to be spherically complete and to have a valuation group equal to $\mathbb{R}$.

However such a conjecture is far from beeing assured. Indeed the process used in the proof of Theorem 4 explicitly will involve the diameter of $F$. On the other hand, if a $T$-filter has no center in $K$ we don't see any process to reduce the problem. Actually we don't know a lot of functions $\sqrt[q]{f}$ which are not analytic element (while $f$ is so) except the functions $\sqrt[q]{1+x}$ or something of that kind...

## §2. MOTZKIN'S FACTORIZATION AND PROOF OF THEOREMS 1, 2, 3.

Let $D$ be a closed bounded infraconnected set, let $T$ be a hole of $D$ and let $a \in T$. Let $f$ be a semi-invertible element in $H(D)$. By Motzkin's Factorization Theorem [M], $f$ factorizes in a unique way in the form $\widehat{f}^T(x-a)^{m(f,T)} f_a^T$ with $\widehat{f}^T \in H(D \cup T)$, and $\widehat{f}^T(x) \neq 0$ whenever $x \in T$, $m(f,T) \in \mathbb{Z}$, and $m(f,T)$ does not depend on the point $a \in T$, $f_a^T \in H(K \backslash T)$, $f_a^T(x) \neq 0$ whenever $x \in K \backslash T$ and $\lim_{|x| \to \infty} f_a^T(x) = 1$.

The following Lemma is then an obvious but helpful remark.

**LEMMA 1.** *Let $D$ be a closed bounded infraconnected set and let $f, g$ be semi-invertible elements in $H(D)$. Then $(\widehat{fg})^T = \widehat{f}^T \cdot \widehat{g}^T$, $(f_a g_a)^T = f_a^T \cdot g_a^T$ and $m(fg,T) = m(f,T) + m(g,T)$.*

LEMMA 2. *Let $D$ be a closed bounded infraconnected set, let $T$ be a hole of $D$ and let $f$ be a semi-invertible element in $H(D)$. If $f$ has continuation to an element of $H(D \cup T)$ then $f$ is semi-invertible in $H(D \cup T)$.*

*Proof.* By hypothesis $f$ factorizes in $H(D)$ in the form $P(x)g(x)$ with $P$ a polynomial whose zeros are in $D$ and $g$ an invertible element in $H(D)$. Let $T = d(a, r^-)$, let $a_1, \ldots, a_n$ be the zeros of $P$, let $\rho \in ]0, r[$ and let $D'_\rho = D \cup T \setminus \left( \bigcup_{i=1}^{n} d(a_i, \rho^-) \right)$. Then $\frac{1}{P} \in H(D'_\rho)$, $f \in H(D'_\rho)$ hence $g \in H(D'_\rho)$. This is true for every $\rho > 0$ and $g$ is bounded in $D$ hence we know that for every $i = 1, \ldots, n$ the Mittag-Leffler term of $g$ in the hole $d(a_i, \rho^-)$ is zero hence finally $g \in H(D \cup T)$.

LEMMA 3. *Let $D$ be a closed bounded infraconnected set and let $T$ be a hole of $D$. Let $f$ be a semi-invertible element in $H(D)$ which has continuation to an element of $H(D \cup T)$. Let $Q$ be the monic polynomial whose zeros are those of $f$ in $T$ (taking multiplicities into account). Let $a \in T$. Then as an element of $H(D)$, $f$ satisfies $m(f,T) = \deg Q$ and $f_a^T = \frac{Q(x)}{(x-a)^{m(f,T)}}$.*

*Proof.* By Lemma 2 $f$ is semi-invertible in $H(D \cup T)$ hence it factorizes in the form $P(x)g(x)$ with $P$ a monic polynomial whose zeros are in $D \cup T$ and $g$ an invertible element in $H(D \cup T)$.

Now $Q$ clearly divides $P$ in $K[x]$ and $P(x) = R(x)Q(x)$. Let $t = \deg Q$; then $\frac{Q(x)}{(x-a)^t}$ belongs to $H(K \setminus T)$; let $h$ be this rational function; clearly we have $\lim_{|x| \to +\infty} h(x) = 1$. Finally by uniqueness of Motzkin's Factorization (once $a$ is a fixed in $T$) we see that $\widehat{f}^T(x) = R(x)f(x)$, $f^T(x) = \frac{Q(x)}{(x-a)^t}$ and $m(f,T) = t$.

*Proof of Theorem* 1. Let $a \in T$. On the first hand $f$ factorizes in $H(D)$ in the form $\widehat{f}^T(x-a)^t f_a^T$ with $t = m(f,T)$.

On the second hand, let $Q$ be the monic polynomial whose zeros are the zeros of $f^s$ in $T$ (taking multiplicities into account). By Lemma 1 and 3 we have $\widehat{f^s}^T = \left(\widehat{f}^T\right)^s, (f^s)_a^T = \left(f_a^T\right)^s = \frac{Q(x)}{(x-a)^{ts}}$ hence $\deg Q = ts$ and then the number of zeros of $f^s$ in $T$ is clearly a multiple of $s$.

Let us first suppose that $\deg Q = 0$, hence $m(f,T) = 0$. Then $\left(f_a^T\right)^s = 1$ hence by $\lim_{|x| \to \infty} f_a^T(x) = 1$ we see that $f_a^T = 1$, hence $f \in H(D \cup T)$, and that finishes proving Theorem 1.

*Proof of Theorem* 2. Since $f \notin H(d(0,r))$ we certainly may assume $r \in |K|$. The Mittag-Leffler series of $f$ on $d(0, r^-)$ is in the form $\sum_{n=0}^{\infty} f_n$ with $f_0 \in d(0,r)$ and

for each $n \in \mathbf{N}^*$, $f_n \in H_0(K \backslash d(a_n, r^-))$ with $|a_n - a_m| = |a_n| = r$ whenever $n \neq m$. We put $T_n = d(a_n, r^-)$. Since $f \notin H(d(0,r))$ at least one of the $f_n$ is different from zero. Let $\ell$ be an integer such that $f_\ell \neq 0$. Then $f$ has no continuation to an element of $H(d(0,r^-) \cup T_\ell)$. Now since $f^s \in H(d(0,r))$, by Theorem 1, $f^s$ has a number of zeros in $T_\ell$ which is different from zero and multiple of $s$. Since the number of zeros of an analytic element in $d(0,r)$ is finite, we see there are finitely many integers $\ell$ such that $f_\ell \neq 0$; let $I$ be the finite set of the $n \in \mathbf{N}$ such that $f_n \neq 0$, and let $D = d(0,r) \backslash \left( \bigcup_{\ell \in I} d(a_\ell, r^-) \right)$.
Then by definition $f \in H(D)$ and for every $\ell \in I$ the number of zeros of $f^s$ in $T_\ell$ is different from zero and multiple of $s$. Theorem 2 is then clearly proven.

*Proof of Corollary* 1. Since $r' > r$ obviously $s \geq 2$. Let us suppose that $f \in H(d(0,r^-))$. Since $r' > r$, $f^s$ belongs to $H(d(0,r)$ hence its number of zeros in $C(0,r)$ is different from zero and multiple of $s$ which contradicts the hypotheses. Hence $f \notin H(d(0,r^-))$.

*Proof of Theorem* 3. First let us check that $\psi_q \in A(d(0,1^-))$.

If $K$ has characteristic 0, we know the power series $g(x) = \sum_{n=0}^{\infty} \binom{\frac{1}{q}}{n} x^n$ has a convergence radius equal to $1[K0]$ and the series obviously defines $\psi_q(x)$ because $g(x) \in d(1,1^-)$.

Now let us suppose $K$ to have a characteristic different from zero. Then $K$ has a characteristic equal to its residue characteristic $p$, and is an extension of $\mathbf{F}_p$. The function $\psi_q$ is now defined as follows.

For every $t \in \mathbf{N}$. Let $\phi_t$ be defined in $\mathbf{Z}$ by $\phi_t(x) = \binom{x}{t}$. Then $\phi_t$ is uniformly continuous in $\mathbf{Z}$ with respect to the $p$-adic value of $\mathbf{Z}$, and then $\phi_t$ has continuation to a continuous mapping from $\mathbf{Z}_p$ into $\mathbf{Z}_p$.

Now $\mathbf{F}_p$ is isomorphic to the residue class field of $\mathbf{Z}_p$ by its maximal ideal $p\mathbf{Z}_p$; let $\overline{x}$ denote the class of $x$ in $\mathbf{F}_p (x \in \mathbf{Z}_p)$. Then for every $y \in \mathbf{F}_p$ and $n \in \mathbf{N}$ and $x \in \mathbf{Z}_p$ such that $\overline{x} = y$, we may define $\binom{y}{n} = \overline{\binom{x}{n}}$. In particular we may define $\binom{\frac{1}{\overline{q}}}{n} = \overline{\binom{\frac{1}{q}}{n}}$. Then the series $g(x) = \sum_{n=0}^{\infty} \binom{\frac{1}{q}}{n} x^n$ satisfies $(g(x))^q = 1 + x$ while $g(x) \in d(1,1^-)$ hence $g(x) = \psi_q(x)$. On the other hand $\psi_q$ has a radius of convergence equal to 1 because the absolute value of $K$ induces the trivial absolute value on $\mathbf{F}_p$. That finishes showing that $\psi_q \in \mathbf{A}(d(0,1^-))$.

Then $(\psi_q(x))^q$ is the polynomial $x+1$, hence has only one zero in $C(0,1)$, and then by Corollary 1 $\psi_q \notin H(d(0,1^-))$. That ends the proof of Theorem 3.

## §3. THE VALUATION FUNCTION AND THE MITTAG-LEFFLER THEOREM

Let $p \in ]1, +\infty[$, and let log be the logarithm function of base $p$ from $\mathbb{R}_+^*$ to $\mathbb{R}$. We put $v(x) = -\log|x|$ for every $x \in K^*$.

Let $D$ be an infraconnected set of diameter $R$ and let $a \in \widetilde{D}$. If $a \in \overline{D}$ let $\mu \in [-\log R, +\infty[$ and if $a$ belongs to a hole of diameter $r$, let $\mu \in [-\log R, -\log r]$. We know that for $h \in H(D)$, $v(h(x))$ admits a limit $\lambda \in \overline{\mathbb{R}}$ when $v(x-a)$ approaches $\mu$ while being different from $\mu$, and while $x$ belongs to $D$ [E3]. Moreover if $D$ has no $T$-filter then $\lambda \in \mathbb{R}$. We denote by $v_a(h, \mu)$ this limit $\lambda$ [E1,E2,E3,K2,R]. Then the function $\mu \to v_a(h, \mu)$ is continuous and piecewise linear in $[-\log R, +\infty[$. Moreover, when $D$ has no $T$-filter we know that for every $r \in ]0, R]$ $h$ satisfies $v(h(x)) = v_a(h, v(x-a))$ for every $x \in D \cap \Delta(a, r, R)$ except maybe in finitely many classes of finitely many circles of center $a : C(a, s_i)_{1 \le i \le q}$ with $s_i \in [r, R]$, [E3,E4].

Suppose now $D$ has no $T$-filter, let $u, w \in H(D)$, and let $Z$ be the set of zeros of $w$ in $D$; in the same way as in [E6] we can define a valuation function for $\frac{u}{w}$ by putting

$$v_a\left(\frac{u}{w}, \mu\right) = v_a(u, \mu) - v_a(w, \mu).$$

The function $\mu \mapsto v_a\left(\frac{u}{w}, \mu\right)$ is clearly continuous and piecewise linear in $[-\log R, +\infty[$ and for every $r \in ]0, R[$ the equality $v\left(\frac{u}{w}(x)\right) = v_a\left(\frac{u}{w}, v(x-a)\right)$ holds in all of $(D \setminus Z) \cap \Delta(a, r, R)$ except maybe in finitely many classes of finitely many circles $C(a, s_i)_{1 \le i \le q}$ with $r \le s_i \le R$.

Let $f \in H(D)$. The Mittag-Leffler Theorem [K1,K2,R] shows that:

There exists a unique sequence of holes $(T_n)_{n \in \mathbb{N}^*}$ of $D$ together with a unique sequence $f_n \in H(K \setminus T_n)$, $f_n \not\equiv 0$, such that $\lim_{|x| \to \infty} f_n(x) = 0$ (whenever $n \in \mathbb{N}^*$) and there exists a unique $f_0 \in H(d(a, R))$ such that the series $\sum_{n=0}^{\infty} f_n$ converges to $f$ in $H(D)$.

Moreover we have: $\|f_0\|_D = \|f_0\|_{d(a,R)}$ and $\|f_n\|_D = \|f_n\|_{K \setminus T_n}$ whenever $n \in \mathbb{N}^*$ and $\|f\|_D = \max(\|f_0\|_D, \sup_{n \in \mathbb{N}^*} \|f_n\|_{K \setminus T_n})$.

Every hole $T_n$ is called a *$f$-hole* and $f_n$ will be called *the Mittag-Leffler term of $f$ associated to the hole $T_n$*.

If $T = d(a, r^-)$ is a $f$-hole of $D$ and if $g$ is the Mittag-Leffler term of $f$ associated to $T$ we also know that $-\log \|g\|_D = v_a(g, -\log r)$.

Let $D$ be an infraconnected set, let $a \in \widetilde{D}$, let $\rho$ be the distance from $a$ to $D$, and let $R$ be the diameter of $D$. Let $r \in ]\rho, R] \cap \mathbb{R}_+^*$. We call an *increasing filter of center $a$, of diameter $r$* the $T$-filter $F$ which admits for base the sets $\Gamma(a, s, r) \cap D (\rho < s < r)$; $F$ is said *to be pierced* if for every $s \in ]\rho, r[$ $\Gamma(a, s, r)$ contains holes of $D$.

$P(F)$ denotes $D \backslash d(a, r^-)$, called *the beach* of $F$

The increasing pierced filter $F$ is called a *$T$-filter* if the holes of $D$ included in the sets $\Gamma(a, s, r)$ satisfy a condition $(*)$ given in many previous articles [E3,E4,R]. We won't explicitly use it in the proof of the Theorems.

Similarly let $r \in [\rho, R[ \cap \mathbb{R}_+^*$. We call *a decreasing filter of center $a$, of diameter $r$* the filter $F$ which admits for base the sets $\Gamma(a, r, s) \cap D$ $(r < s < R)$; $F$ is said *to be pierced* if for every $s \in ]r, R[$ $\Gamma(a, r, s)$ contains holes of $D$.

$P(F)$ denotes $d(a, r) \cap D$, called *the beach* of $F$.

The decreasing pierced filter is called a $T$-filter if the holes of $D$ included in the sets $\Gamma(a, r, s)$ satisfy the same condition $(*)$ as above.

When $K$ is not spherically complete we also have to define decreasing filter with no center. Let $d(a_n, r_n)_{n \in \mathbb{N}}$ be a sequence of disks included in $\widetilde{D}$, such that

$$d(a_n, r_n) \cap D \neq \emptyset$$

$$d(a_{n+1}, r_{n+1}) \subset d(a_n, r_n)$$

$$\lim_{n \to \infty} r_n = r > 0$$

$$\bigcap_{n \in \mathbb{N}} d(a_n, r_n) = \emptyset.$$

We put $D_n = d(a_n, r_n) \cap D$ $(n \in \mathbb{N})$ and the filter $F$ which admits for base the sequence $D_n$ is called *a decreasing filter with no center. $F$* is said *to be pierced if for* every $n \in \mathbb{N}$, $d(a_n, r_n)$ contains holes of $D$, and $F$ is called a *decreasing $T$-filter with no center* if the holes of $D$ included in the $d(a_n, r_n)$ satisfy the condition $(*)$ again.

## §4. PROOF OF THEOREM 4

We will first show the following Lemma.

**LEMMA L.** *Let $D$ be an infraconnected set with an increasing T-filter $F$ of center 0, of diameter $r$ and let $R$ be such that $r \leq R \leq \operatorname{diam}(D)$.*

*Let $\pi \in H(D)$ be such that $\lim_F \pi(x) = 0$ and $\pi(x) = 0$ whenever $x \in P(F)$ and let $f$ be a bounded power series convergent in $d(0, R^-)$. Then the function $h$ defined in $D$ by $h(x) = \pi(x)f(x)$ whenever $x \in D \cap d(0, R^-)$ and $h(x) = 0$ whenever $x \in D \backslash d(0, R^-)$ belongs to $H(D)$.*

*Proof of Lemma L.* Let $f(x) = \sum a_n x^n$, let $M = \|f\|_{d(0,R^-)}$ and for every $n \in \mathbb{N}$ let $f_n = \sum_{i=0}^{n} a_i x^i$. Let us fix $\varepsilon \in ]0, 1[$. There exists $\rho(\varepsilon) \in ]0, R^-[$ such that

$$(1) \qquad \left|\pi(x)\right| \leq \frac{\varepsilon}{M} \text{ whenever } x \in D \backslash d(0, \rho(\varepsilon)).$$

Now let $B = \max(\|\pi\|_D, 1)$ and let $N(\varepsilon) \in \mathbb{N}$ be such that $\|f - f_n\|_{d(0,\rho(\varepsilon))} \leq \frac{\varepsilon}{B} < M$ for all $n \geq N(\varepsilon)$. Then let us take $n \geq N(\varepsilon)$ we have $|\pi(x)f(x) - \pi(x)f_n(x)| \leq \varepsilon$ for all $x \in d(0, \rho(\varepsilon)) \cap D$ and by (1) we see that $|\pi(x)f(x) - \pi(x)f_n(x)| \leq \frac{\varepsilon}{M}|f(x) - f_n(x)|$. But by definition of $M$ we have $|a_i|\ |x|^i \leq M$ for every $i \in \mathbb{N}$ hence $\|f_n\|_{d(0,R^-)} \leq M$ hence $|f(x) - f_n(x)| \leq M$ and therefore we have (2) $|\pi(x)f(x) - \pi(x)f_n(x)| \leq \varepsilon$ for all $x \in d(0, R^-) \backslash d(0, \rho(\varepsilon))$.

At last we just have (3) $\pi(x)f(x) - \pi(x)f_n(x) = 0$ for all $x \in D \backslash d(0, R^-)$.

Thus by (1), (2), (3) we see that $\|\pi f - \pi f_n\|_D \leq \varepsilon$ and that just proves that $\pi f$ belongs to $H(D)$.

*Proof of Theorem 4.* We will first suppose $F$ to be increasing. Let $a$ be a center and let $r$ be its diameter. Thanks to the fact that one of the conditions $\alpha$) and $\beta$) is satisfied, it is easily seen that there exists $R \in |K|$ such that $r \leq R \leq \operatorname{diam}(D)$. Then there exists a mapping $\phi$ of the form $\phi(x) = \gamma x + \delta$ such that $\phi(a) = 0$, and $\phi(d(a, R)) = d(0, 1)$. Let $D' = \phi(D)$; the algebras $H(D)$ and $H(D')$ are isomorphic so that we can here assume $a = 0$ and $R = 1$. We then have $r \leq 1 \leq \operatorname{diam}(D)$.

Let $B = d(0, 1^-)$ and let $q$ be an integer superior or equal to 2, prime to the residue characteristic $p$ of $K$ in the case when $p \neq 0$. By Theorem 3 we know that $\psi_q \in \mathbf{A}(B) \backslash H(B)$.

Let $D'' = D \cap B$. We can easily check that $\psi_q$ does not belong to $H(D'')$. Indeed suppose it does. Since $\psi_q$ is bounded in $B$ it obviously belongs to $H(\overline{D''})$. The Mittag-Leffler series of $f$ on $\overline{D''}$ has no term associated to any one of the holes $T$ of $\overline{D''}$ included in $B$ because $\psi_q$ obviously belongs to $H(d(0, \rho))$ for every $\rho \in ]0, 1[$ and therefore $\psi_q \in H(B)$ which is false and that finishes the proof of $\psi_q \notin H(D'')$.

Let $\pi \in H(D)$ be such that $\lim_F \pi(x) = 0$ and $\pi(x) = 0$ whenever $x \in P(F)$.

By Lemma L, the function $h$ defined by $h(x) = \pi(x)\psi_q(x)$ whenever $x \in D \cap B$ and $h(x) = 0$ whenever $x \in D \backslash B$ belongs to $H(D)$, hence there exists $\widehat{\psi}_q$ in the quotients field of $H(D)$ such that $\widehat{\psi}_q(x) = \psi_q(x)$ for every $x \in D \cap B$.

On the other hand it is easily seen that $\widehat{\psi}_q$ is integral over $H(D)$.

The relation $h^q = (1+x)\pi^q$ is obvious when $x \in D \cap B$ because $(\psi_q(x))^q = x+1$, and the relation is trivial in $D \backslash B$ because $h(x) = \pi(x) = 0$. Thus we have $\left(\frac{h}{\pi}\right)^q = 1+x$ in the quotients field of $H(D)$.

But then, if $H(D)$ is integrally closed we must have $\widehat{\psi}_q \in H(D)$, and in particular $\psi_q \in H(D'')$ which is false. So we have arrived at a contradiction and proven that $H(D)$ is not integrally closed when $F$ is increasing.

Suppose now $F$ is decreasing of center $a$, of diameter $r$. If $d(a,r)$ contains at least a point $a \notin D$, an inversion $\psi$ of center $a$ maps $D$ onto a set $D'$ with an increasing $T$-filter $F'$ and then $H(D)$ is isometric to $H(D')$. By what has been seen above $H(D')$ is not integrally closed hence $H(D)$ is not either.

Now suppose that $d(a,r) \subset D$. An inversion $\psi$ of center $a$ maps $D \backslash \{a\}$ onto a set $D'$ and then $H(D)$ is isomorphic to the subalgebra $A$ of elements $f \in H(D')$ which admit a limit when $|x|$ tends to $+\infty$. In the construction made in the case when there is an increasing $T$-filter we saw that the elements $\pi$ and $h$ do belong to $A$, and satisfy $(h(x))^q = (1+x)(\pi(x))^q$ in all of $D'$ hence the relation $h^q = (1+x)\pi^q$ is true in $A$. If $H(D)$ is integrally closed, so is $A$ hence $\frac{h}{\pi} \in A$. But we saw above that $\frac{h}{\pi} \notin H(D')$ hence $\frac{h}{\pi} \notin A$. That ends the proof of Theorem 4.

## REFERENCES

[A] Amice, Y.: *Les nombres p-adiques.* P.U.F. (1975).

[E1] Escassut, A.: *Les algèbres de Krasner.* C.R. Acad. Sci. Paris Sér. A-B **272** (1971), A598–A601.

[E2] Escassut, A.: *Algèbres d'éléments analytiques en analyse non archimédienne,* Nederl. Akad. Wetensch. Proc. Ser. A **77** = Indag. Math. **36** (1974),339–351.

[E3] Escassut, A.: *Eléments analytiques et filtres percés sur un ensemble infraconnexe.* Ann. Mat. Pura Appl. (4) **110** (1976), 335–352.

[$E_4$] Escassut, A.: *T-filtres, ensembles analytiques et transformation de Fourier p-adique.* Ann. Inst. Fourier (Grenoble) **25** (1975), 45–80.

[$E_5$] Escassut, A.: *Algèbres de Krasner intègres et noethériennes.* Nederl. Akad. Wetensch. Proc. Ser. A **79** = Indag. Math. **38** (1976), 109–130.

[$E_6$] Escassut, A.: *Integrally Closed Algebras of Analytic Elements.* To appear in Communications in Algebra.

[G] Garandel, G.: *Les semi-normes multiplicatives sur les algèbres d'éléments analytiques au sens de Krasner.* Nederl. Akad. Wetensch. Proc. Ser. A **78** = Indag. Math. **37** (1975), 327–341.

[KO] Koblitz, N.: *p-Adic numbers, p-adic Analysis and Zeta Function.* Springer-Verlag, New York-Heidelberg-Berlin, 1977.

[$K_1$] Krasner, M.: *Prolongement analytique dans les corps valués complets: préservation de l'analycité par la convergence uniforme et par la dérivation; théorème de Mittag-Leffler généralisé pour les éléments analytiques,* C.R. Acad. Sci. Paris **244** (1957), 2570–2573.

[$K_2$] Krasner, M.: *Prolongement analytique uniforme et multiforme dans les corps valués complets.* In *Les tendances géométriques en algèbre et théorie des nombres.* Clermont-Ferrand 1964, 91–141. Centre Nationale de la Recherche Scientifique (1966), (Colloques internationaux du C.N.R.S. Paris, 143).

[M] Motzkin, E.: *La décomposition d'un élément analytique en facteurs singuliers.* Ann. Inst. Fourier (Grenoble) **27** (1977), 67–82.

[R] Robba, Ph.: *Fonctions analytiques sur les corps valués ultramétriques complets. Prolongement analytique et algèbres de Banach ultramétriques.* Astérisque, **10** (1973), 109–220.

[$S_1$] Sarmant, M.C.: *Factorisation de Motzkin en Produit Méromorphe d'un élément semi-inversible.* To appear in the Bulletin des Sciences Mathématiques.

[$S_2$] Sarmant, M.C., Escassut, A.: *The Differential Equation $y' = \omega y$ and the Meromorphic Products.* In *p-adic Functional Analysis*, edited by J.M. Bayod, N. De Grande-De Kimpe and J. Martínez-Maurica, Marcel Dekker, New York (1991), 157–175.

UNIVERSITÉ BLAISE PASCAL
MATHÉMATIQUES PURES
F-63177 AUBIERE CEDEX, FRANCE

# CONTINUOUS OPERATORS WHICH COMMUTE WITH TRANSLATIONS, ON THE SPACE OF CONTINUOUS FUNCTIONS ON $\mathbb{Z}_p$

Lucien van Hamme

## 1. INTRODUCTION

Let $p$ be a prime number. $\mathbb{Z}_p$ and $\mathbb{Q}_p$ denote, respectively, the ring of $p$-adic integers and the field of $p$-adic numbers. $K$ is a non-archimedean non trivially valued complete field with valuation $|\cdot|$. We suppose that $\mathbb{Q}_p \subset K$. The space of all continuous functions $f : \mathbb{Z}_p \to K$, equipped with the supremum norm $\|\cdot\|$, will be denoted by $C(\mathbb{Z}_p \to K)$. The degree of a polynomial $p$ will be denoted deg $p$.

We are interested in linear continuous operators on $C(\mathbb{Z}_p \to K)$.
Examples of such operators are:

- the translation operator $E$ which transforms $f$ in $Ef$ defined by $(Ef)(x) = f(x+1)$;
- the general translation operator $E^\alpha$ defined by $(E^\alpha f)(x) = f(x+\alpha)$, $\alpha \in \mathbb{Z}_p$;
- the difference operator $\Delta$ with $(\Delta f)(x) = f(x+1) - f(x)$.

The purpose of this paper is to study the linear continuous operators on $C(\mathbb{Z}_p \to K)$ which commute with $E$.
Let $Q$ be such an operator. Since the binomial polynomials $B_n(x) = \binom{x}{n}$, $n = 0, 1, \ldots$ form a base of $C(\mathbb{Z}_p \to K)$ it is natural to look at $QB_n$. Put $b_n = (QB_n)(0)$. In section 2 of this paper we will prove that the operators $Q$ can be characterized in the following way.

**PROPOSITION.** *A linear continuous operator on $C(Z_p \to K)$ commutes with $E$ if, and only if, the sequence $(b_n)$ is bounded.*

The main result of this paper is:

**THEOREM.** *If $Q$ is a linear continuous operator on $C(\mathbf{Z}_p \to K)$ such that $b_0 = 0$, $|b_1| = 1$, $|b_n| \leq 1$ for $n \geq 2$ then*

a) *there exists a unique sequence of polynomials $q_n(x)$ such that*

$$Qq_n = q_{n-1},\ deg\ q_n = n,\ q_n(0) = 0\ for\ n \geq 1\ and\ q_0 = 1.$$

b) *every continuous function $f : \mathbf{Z}_p \to K$ has a uniformly convergent expansion of the form*

$$f(x) = \sum_{n=0}^{\infty} (Q^n f)(0) q_n(x), \qquad \|f\| = \max_n |(Q^n f)(0)|.$$

## Remarks

1) The $q_n(x)$ form a base of $C(\mathbf{Z}_p \to K)$. Many other bases are known (see e.g. [3]). What makes the theorem interesting is that we not only have a base but also an expression for the coefficients of $q_n(x)$.

2) The polynomials $q_n(x)$ have been studied by G.C. Rota [2], S. Roman [1], and many others. They play an essential rôle in the so-called "umbral calculus". In these papers the authors work over the complex field while the operators usually act on the space of (complex) polynomials. The expansion of the theorem then reduces to a finite polynomial identity.

We finish this introduction by giving two examples. In each example we give $Q$ and the corresponding polynomials $q_n(x)$. Since these polynomials are unique it is sufficient to verify the conditions on $b_n$ and the conditions (a) of the theorem. Other examples can be found in section 4.

**EXAMPLE 1.** If $Q = \Delta$ then $b_1 = 1$ and $b_n = 0$ for $n \neq 1$.
The binomial polynomials $q_n(x) = \binom{x}{n}$ verify conditions (a) and we get Mahler's expansion

$$f(x) = \sum_{n=0}^{\infty} (\Delta^n f)(0) \binom{x}{n}$$

**EXAMPLE 2.** $Q = E^\alpha \Delta$ with $\alpha \in \mathbf{Z}_p$. Here $b_0 = 0$, $b_1 = 1$. For $n \geq 2$ $b_n = \binom{\alpha}{n-1}$ hence $|b_n| \leq 1$.

In this case $q_n(x) = \frac{x}{n}\binom{x-1-\alpha n}{n-1}$, $n \geq 1$.
In order to verify that $Qq_n = q_{n-1}$ we write $q_n(x)$ in the form

$$q_n(x) = \binom{x-1-\alpha n}{n} + (\alpha+1)\binom{x-1-\alpha n}{n-1}.$$

For the first term we see that

$$E^\alpha \Delta\binom{x-1-\alpha n}{n} = E^\alpha\binom{x-1-\alpha n}{n-1} = \binom{x-1-\alpha(n-1)}{n-1}.$$

Since we can write a similar result for the second term we have that $Qq_n = q_{n-1}$ and the theorem gives

$$f(x) = f(0) + x\sum_{n=1}^{\infty} \frac{(\Delta^n f)(\alpha n)}{n}\binom{x-1-\alpha n}{n-1}.$$

This is theorem 2 of [4].

## 2. OPERATORS COMMUTING WITH THE TRANSLATION OPERATOR

Our first aim in this section is to prove the proposition from the introduction.
Let $(b_n)$ be a bounded sequence in $K$ and let $f$ be an element of $C(\mathbb{Z}_p \to K)$.
We define an operator $Q$ on $C(\mathbb{Z}_p \to K)$ by

$$(Qf)(x) = \sum_{n=0}^{\infty} b_n(\Delta^n f)(x). \tag{1}$$

Since $\lim_{n\to\infty}(\Delta^n f)(x) = 0$ uniformly, the R.H.S. of (1) converges uniformly and defines a continuous function $Qf$. The operator $Q$ is clearly linear and commutes with $E$ since $E\Delta = \Delta E$. $Q$ is continuous since $\|Qf\| \leq \|f\| \cdot (\max_n |b_n|)$.
The numbers $b_n$ have the meaning of the introduction i.e. $b_n = (QB_n)(0)$.

In order to prove the proposition we have to show that, conversely, for every operator $Q$ which commutes with $E$ there is a bounded sequence such that (1) holds.
We need three lemmas.

LEMMA 1. *If $f : \mathbb{Z}_p \to K$ is a continuous function and if $\Delta^{n+1}f = 0$ then $f$ is a polynomial of degree not greater than $n$.*

*Proof.* By induction.
If $\Delta f = 0$ then $f(x+1) = f(x)$ for all $x$ in $\mathbf{Z}_p$. Hence $f(x)$ is constant which proves the case $n = 0$. Suppose that the lemma is true with $n$ replaced by $(n-1)$ and let $f$ be a continuous function such that $\Delta^{n+1} f = \Delta^n(\Delta f) = 0$.
It follows from the induction hypothesis that $\Delta f$ can be written in the form

$$(\Delta f)(x) = \sum_{k=0}^{n-1} a_k \binom{x}{k}. \qquad (*)$$

The polynomials of the form $f(x) = \sum_{k=0}^{n-1} a_k \binom{x}{k+1} + c$ satisfy $(*)$. They are the only continuous functions which satisfy $(*)$ for if $\Delta f = \Delta g$ then $f(x) = g(x)+c$. Hence $f$ is a polynomial of degree not greater than $n$. This proves the lemma.

LEMMA 2. *Let $p$ be a polynomial of degree $n$ with coefficients in $K$ and let $Q$ be an operator such that $EQ = QE$. Then $Qp$ is a polynomial of degree not greater than $n$.*

*Proof.* Since $QE = EQ$ implies that $Q\Delta = \Delta Q$ we see that $Q\Delta^{n+1} = \Delta^{n+1} Q$. Since $\deg p = n$ we have $\Delta^{n+1} p = 0$ and hence $\Delta^{n+1}(Qp) = Q(\Delta^{n+1} p) = 0$. Using lemma 1 we see that $Qp$ is a polynomial of degree not greater than $n$.

Lemma 2 says that $Q$ does not increase the degree of a polynomial. We now show that $Q$ lowers the degree of all polynomials which are not in the kernel of $Q$ with the same amount. As always $Q$ is a continuous linear operator on $C(\mathbf{Z}_p \to K)$.

LEMMA 3. *If $QE = EQ$ then the integer $m = \deg p - \deg(Qp)$ is the same for all $p$ in $K[x]$ which are not in the kernel of $Q$.*

*Proof.* We examine the kernel of $Q$. Suppose that $p$ is a polynomial of Ker $Q$ i.e. $Qp = 0$.
Since $Q(\Delta p) = \Delta(Qp) = 0$ we see that $\Delta p$ also belongs to Ker $Q$. In the same way we can conclude that $\Delta^2 p, \Delta^3 p, \ldots$ belong to Ker $Q$. Using the linearity of $Q$ we see that Ker $Q$ contains all polynomials of degree $\leq \deg p$.
If Ker $Q$ contains polynomials of arbitrarily high degree then $K[x] \subset$ Ker $Q$ and there is nothing to prove.
Hence we may assume that Ker $Q$ consists of all polynomials of degree strictly smaller than an integer $m$. Take a polynomial $p$ with $\deg p = n \geq m$. We claim that deg $(Qp)$ is $(n-m)$.

Since deg $(\Delta^{n-m+1}p)$ is $(m-1)$ we have $Q(\Delta^{n-m+1}p) = 0$ and hence $\Delta^{n-m+1}(Qp) = 0$. We conclude from lemma 1 that deg $(Qp) \leq n-m$.
The degree of $Qp$ cannot be strictly smaller than $n-m$, for if deg $(Qp) = r < n-m$ we would have $\Delta^{r+1}(Qp) = 0$ and $Q(\Delta^{r+1}p) = 0$.
But deg $(\Delta^{r+1}p) = n-r-1 > m-1$ and Ker $Q$ would contain a polynomial of degree greater than $(m-1)$. Since this contradicts the definition of $m$ we conclude that $\deg(Qp) = n-m$.

COROLLARY. *If* Ker $Q$ *contains* $B_{m-1}$ *then* $Q$ *lowers the degree of every polynomial with at least* $m$.

We can now finish the proof of the proposition.
Let $Q$ be a continuous operator on $C(\mathbb{Z}_p \to K)$ such that $QE = EQ$.
We will construct a bounded sequence $(b_n)$ and we will show that (1) holds.
Let $I$ be the identical operator $C(Z_p \to K)$. Define $b_0$ by $QB_0 = b_0$.
Then Ker $(Q-b_0I)$ contains $B_0$ and by the corollary of lemma 3 we know that $(Q-b_0I)$ lowers the degree of every polynomial with at least 1. So $(Q-b_0I)B_1$ is a constant and we can define $b_1$ by $(Q-b_0I)B_1 = b_1$. Hence Ker $(Q-b_0I)B_1$ is a constant and we can define $b_1$ by $(Q-b_0I)B_1 = b_1$. Hence Ker $(Q-b_0I-b_1\Delta)$ contains $B_1$ etc...
If $b_0, b_1, \ldots b_{n-1}$ have already been defined we see that $(Q-b_0I-\sum_{i=1}^{n-1} b_i\Delta^i)$ contains $B_{n-1}$ and we deduce from the corollary of lemma 3 that $(Q-b_0I-\sum_{i=1}^{n-1} b_i\Delta^i)B_n$ is a constant, so we can define $b_n$ by

$$b_n = (Q-b_0I-\sum_{i=1}^{n-1} b_i\Delta^i)B_n.$$

Noticing that $\Delta B_n = B_{n-1}$ or, more generally, $(\Delta^i B_n)(x) = \binom{x}{n-i}$ we can write this in the form

$$b_n = Q\binom{x}{n} - \sum_{i=0}^{n-1} b_i\binom{x}{n-i}.$$

We will now prove that $(b_n)$ is bounded. Denoting the norm of the (continuous) operator $Q$ by $\|Q\|$ we have $\|Qf\| \leq \|Q\| \cdot \|f\|$.
Hence $|b_n| \leq \max(\|Q\|, |b_0|, |b_1|, \ldots, |b_{n-1}|)$.
Since $|b_0| = \|QB_0\| \leq \|Q\|$ we conclude, by induction on $n$, that $|b_n| \leq \|Q\|$.

It is clear from the construction of the $b_n$ that the kernel of the operator $Q - b_0 I - \sum_{n=1}^{\infty} b_n \Delta^n$ contains $K[x]$. Since $K[x]$ is dense in $C(\mathbf{Z}_p \to K)$ it is the zero operator and hence $Q = b_0 I + \sum_{n=1}^{\infty} b_n \Delta^n$ and we have proved the proposition.

**Remarks and conclusions**

1) The set of all linear continuous operators on $C(\mathbf{Z}_p \to K)$, such that $EQ = QE$, forms a ring under addition and composition. This ring is isomorphic to the ring of formal power series $\sum_{n=0}^{\infty} b_n t^n$ with bounded coefficients.

2) We will call $Q = b_0 I + \sum_{n=1}^{\infty} b_n \Delta^n$ the $\Delta$-expansion of the operator $Q$. We will usually omit $I$ and simply write $Q = \sum_{n=0}^{\infty} b_n \Delta^n$.

In the next section we will constantly use the fact that operators which commute with $E$ can be treated as formal power series.

3) It is a simple matter to prove that $\|Q\| = \sup |b_n|$. Since we will not use this in the sequel we omit the proof.

4) $EQ = QE$ implies commutativity.

The ring of operators commuting with $E$ can be described in another way.

Let $\mu$ be a measure on $\mathbf{Z}_p$. The set of values $\mu(a+p^n\mathbf{Z}_p)$ is bounded. Hence the following definition makes sense

$$\|\mu\| = \sup |\mu(a+p^n\mathbf{Z}_p)| \qquad a \in \mathbf{Z}_p, n \in \mathbf{N}.$$

Every function $f \in C\,(\mathbf{Z}_p \to K)$ can be integrated with respect to $\mu$ and the following inequality holds

$$\left| \int_{\mathbf{Z}_p} f(t)\mu(t) \right| \leq \|f\| \cdot \|\mu\|. \tag{2}$$

For a given measure $\mu$ we can define an operator on $C(\mathbf{Z}_p \to K)$ by

$$(Qf)(x) = \int_{\mathbf{Z}_p} f(x+t)\mu(t). \tag{3}$$

Using (2) it is easy to verify that $Qf$ is a continuous function.

The operator $Q$ is clearly linear and commutes with $E$.

$Q$ is continuous since $\|Qf\| \leq \|\mu\| \cdot \|f\|$.

In order to find the $\Delta$-expansion of $Q$ we use the Mahler expansion of $f(x+t)$.

$$(Qf)(x) = \int_{\mathbf{Z}_p} f(x+t)\mu(t) = \int_{\mathbf{Z}_p} \left[\sum_{n=0}^{\infty} \Delta^n f(x)\binom{t}{n}\right] \mu(t).$$

Since the Mahler expansion converges uniformly we see that

$$(Qf)(x) = \sum_{n=0}^{\infty} b_n(\Delta^n f)(x) \qquad \text{where } b_n = \int_{\mathbf{Z}_p} \binom{t}{n}\mu(t).$$

Suppose, conversely, that an operator $Q = \sum_{n=0}^{\infty} b_n\Delta^n$ is given, where $(b_n)$ is a bounded sequence. It is well-known that there exists a measure $\mu$ such $b_n = \int_{\mathbf{Z}_p} \binom{t}{n}\mu(t)$. Hence $Q$ can be written in the form (3).

*The Pincherle derivative*

In [2] G.C. Rota introduces the notion of Pincherle derivative. This notion can also be defined in our ring of operators.

Consider the operator $\mathbf{x} : f(x) \to xf(x)$.

This is the only operator in this paper which does not commute with $E$.

Definition: The Pincherle derivative of $Q$ is the operator $Q' = Q\mathbf{x} - \mathbf{x}Q$. In integral form

$$\begin{aligned}(Q'f)(x) &= \int_{\mathbf{Z}_p} (x+t)f(x+t)\mu(t) - x\int_{\mathbf{Z}_p} f(x+t)\mu(t)\\ &= \int_{\mathbf{Z}_p} tf(x+t)\mu(t).\end{aligned}$$

Using this formula for $Q'$ the following properties are easily established

1) If $Q = \sum_{n=0}^{\infty} b_n\Delta^n$ then $Q' = (1+\Delta)\sum_{n=1}^{\infty} nb_n\Delta^{n-1}$
2) $(Q_1+Q_2)' = Q_1' + Q_2'$
3) $(Q_1Q_2)' = Q_1'Q_2 + Q_1Q_2'$

## 3. PROOF OF THE THEOREM

The existence and uniqueness of the polynomials $q_n(x)$ is well-known from classical umbral calculus. Several explicit formulas for $q_n(x)$ can be found in the literature. For instance

$$q_n(x) = \frac{x}{n}{Q'}^{-1}q_{n-1}(x). \tag{4}$$

This is given in theorem 4, p.17 of [2], in a slightly different notation. $Q'$ is the Pincherle derivative.
Our proof relies on the fact that, under the conditions of the theorem, we have $|q_n(x)| \leq 1$. The presence of the denominator $n$ in (4) and in other similar formulas prevents us from deducing the estimate $|q_n(x)| \leq 1$ from it. Therefore we will define $q_n(x)$ in a different way. We add the proof of the uniqueness of the $q_n(x)$ for the sake of completeness.

*Uniqueness of the $q_n$*

Let $(q_i)$ be a sequence of polynomials satisfying the conditions (a) of the theorem and let $p$ be a polynomial of degree $n$. Since the $q_i$ form a base of the space $K[x]$ we can write $p(x)$ in the form

$$p(x) = \sum_{i=0}^{n} c_i q_i(x).$$

Since $Qq_i = q_{i-1}$ we have $Q^r p = \sum_{i=r}^{n} c_i q_{i-r}(x)$.
The condition $q_i(0) = 0$, for $i \geq 1$, implies $(Q^r p)(0) = c_r$ and hence

$$p(x) = \sum_{i=0}^{n} (Q^i p)(0) q_i(x). \tag{5}$$

If $(r_n)$ is another sequence of polynomials satisfying the conditions (a) of the theorem, then

$$r_n(x) = \sum_{i=0}^{n} (Q^i r_n)(0) q_i(x).$$

But since $(Q^i r_n)(0) = 0$ for $i = 0, 1, \ldots, n-1$ and $(Q^n r_n)(0) = 1$ we conclude that $r_n = q_n$.

*Definition of $q_n$*

Let $Q = \sum_{n=1}^{\infty} b_n \Delta^n$ be the given operator. We first define two new operators $R$ and $S$ in the following way.
We write the formal power series $\sum_{n=1}^{\infty} b_n t^n$ in the form

$$\sum_{n=1}^{\infty} b_n t^n = \frac{b_1 t}{\sum_{n=0}^{\infty} a_n t^n}.$$

From the denominator $r(t) = \sum_{n=0}^{\infty} a_n t^n$ we derive a new formal power series:

$$1 - \frac{tr'(t)}{r(t)} = \sum_{n=0}^{\infty} c_n t^n.$$

The operators $R$ and $S$ are defined by $R = \sum_{n=0}^{\infty} a_n \Delta^n$ and $S = \sum_{n=0}^{\infty} c_n \Delta^n$.
Note that the conditions $|b_1| = 1$ and $|b_n| \leq 1$ $(n \geq 2)$ imply that $|a_n| \leq 1$ and $|c_n| \leq 1$. Moreover $a_0 = c_0 = 1$.
Definition: $q_n(x) = b_1^{-n} SR^n \binom{x}{n}, \quad n \geq 1.$
It is now easy to show that these polynomials satisfy the condition (a) of the theorem. Since the $\Delta$-expansion of $R^{-1}$ has bounded coefficients it is legitimate to write $Q$ in the form $Q = b_1 \Delta R^{-1}$ and hence

$$(Qq_n)(x) = b_1^{-(n-1)} \Delta S R^{n-1} \binom{x}{n} = b_1^{-(n-1)} S R^{n-1} \binom{x}{n-1} = q_{n-1}(x).$$

The $\Delta$-expansion $SR^n = 1 + \sum_{i=1}^{\infty} d_i \Delta^i$ starts with a 1. By plugging this into the definition of $q_n$ we see that

$$q_n(x) = b_1^{-n} \left\{ \binom{x}{n} + \sum_{i=1}^{n} d_i \binom{x}{n-1} \right\}. \tag{6}$$

Hence deg $q_n = n$. Formula (6) also shows that $q_n(0) = b_1^{-n} d_n$.
Now $d_n$ is the coefficient of $t^n$ in the expansion of

$$\left\{ 1 - t\frac{\phi'(t)}{\phi(t)} \right\} \phi(t)^n = \phi(t)^n - \frac{t}{n}\frac{d}{dt}[\phi(t)^n].$$

Hence $d_n = 0$ and we have proved that $q_n(0) = 0$.

*End of the proof:*

The inequalities $|a_n| \leq 1$ and $|c_n| \leq 1$ imply that $|d_n| \leq 1$.
From (6) we may conclude that $|q_n(x)| \leq 1$.
Hence the operator $T_x$, defined by $T_x = \sum_{n=0}^{\infty} q_n(x) Q^n$ is continuous.
If we let $T_x$ act on the polynomial $p(y)$ we get $\sum_{n=0}^{m} q_n(x)(Q^n p)(y)$. Comparing this with (5), where we replace $p(x)$ by $p(x+y)$, we see that $(T_x p)(y) = p(x+y)$.

This means that the operators $T_x$ and $E^x$ (the translation operator) coincide on $K[x]$. Since both operators are continuous and since the polynomials are dense in $C\ (\mathbb{Z} \to K)$ they coincide on the whole space i.e.

$$E^x = \sum_{n=0}^{\infty} q_n(x)Q^n.$$

If the operators act on a continuous function $f(y)$ we get

$$f(x+y) = \sum_{n=0}^{\infty} q_n(x)(Q^n f)(y). \tag{7}$$

For $y = 0$ this gives the expansion we wanted to prove.
The formula $\|f\| = \max_n |(Q^n f)(0)|$ now follows easily.

$$|f(x)| = \left|\sum_{n=0}^{\infty}(Q^n f)(0)q_n(x)\right| \leq \max_n |(Q^n f)(0)| \quad \text{since } |q_n(x)| \leq 1.$$

Hence $\|f\| \leq \max_n |(Q^n f)(0)|$.
On the other hand

$$|Q(f)(x)| = \left|\sum_{n=1}^{\infty} b_n(\Delta^n f)(x)\right| \leq \max_n |(\Delta^n f)(x)| \leq \|f\| \quad \text{since } |b_n| \leq 1.$$

Hence $\|Qf\| \leq \|f\|$ and $\|Q^n f\| \leq \|f\|$.
So $|(Q^n f)(0)| \leq \|Q^n f\| \leq \|f\|$ and $\max_n |(Q^n f)(0)| \leq \|f\|$.
This proves the formula for $\|f\|$.

The only thing that we still have not proved is the uniform convergence of (7). This will follow from the inequality $\|(Q^n f)(y)\| \leq \|\Delta^n f\|$ since $\lim_{n\to\infty} \|\Delta^n f\| = 0$.
In order to prove this inequality we notice that we can write $Q^n$ in the form $Q^n = T\Delta^n$ where the $\Delta$-expansion of the operator $T$ is $T = \sum_{i=0}^{\infty} \alpha_i \Delta^i$ with $|\alpha_i| \leq 1$.
Hence $|(Tf)(x)| \leq \|f\|$ and $\|Tf\| \leq \|f\|$.
Finally $|(Q^n f)(y)| \leq \|Q^n f\| = \|T\Delta^n f\| \leq \|\Delta^n f\|$ and the proof of the theorem is complete.

The following result follows immediately from the preceding proof and is a generalization of the theorem.

Let $Q$ be an operator satisfying the conditions of the theorem and let $q_n$ be the associated polynomials.
Let $S$ be an operator commuting with $E$. (Warning: this is not the operator used in the proof above.)
The polynomials $s_n = Sq_n$ are called "Sheffer polynomials".

COROLLARY. *If $S$ is a bijection then*

$$f(x) = \sum_{i=0}^{\infty} (S^{-1}Q^n f)(0) s_n(x), \qquad f \in C(\mathbf{Z} \to K).$$

*Proof.* Consider (7) as an identity in $x$ and let $S$ act on both sides of the identity. This gives

$$(Sf)(x+y) = \sum_{i=0}^{\infty} s_n(x)(Q^n f)(y).$$

Replacing $Sf$ by $g$ and putting $y = 0$ we find

$$g(x) = \sum_{i=0}^{\infty} (S^{-1}Q^n g)(0) s_n(x).$$

**Remark.** With the notations introduced above we also have the following expression for $q_n(x)$:

$$q_n(x) = b_1^n \frac{x}{n} R^n \binom{x-1}{n-1}.$$

This is a particular case of corollary 3.8.2 of [1] p.51.
As stated in the introduction of this section, we can not deduce the inequality $|q_n(x)| \leq 1$ from this. What can be deduced is the inequality $\left|\frac{n}{x} q_n(x)\right| \leq 1$. We will need this in section 4.

## 4. EXAMPLES

EXAMPLE 4.1. Let $Q$ be an operator which satisfies the conditions of the theorem and let $q_n(x)$ be the associated polynomials. The polynomials corresponding to the operator $\overline{Q} = QE^{\alpha}$ are

$$x \frac{q_n(x-n\alpha)}{x-n\alpha}, \qquad n \geq 1.$$

(This is proposition 4, p. 33 of [2]).
If we suppose that $f$ has period $p$, i.e. $f(x+p) = f(x)$ for all $x \in \mathbf{Z}_p$ and if $\alpha \in p\mathbf{Z}_p$ then $\overline{Q}^n f = Q^n f$. The theorem then gives

$$f(x) = \sum_{n=0}^{\infty} (Q^n f)(0) x \frac{q_n(x-n\alpha)}{x-n\alpha}$$

$$\frac{f(x)-f(0)}{x} = \sum_{n=1}^{\infty} (Q^n f)(0) \frac{q_n(x-n\alpha)}{x-n\alpha}, \quad \alpha \in p\mathbf{Z}_p. \tag{8}$$

This R.H.S. is uniformly convergent in $\mathbf{Z}_p$. This can be seen in the following way:

$$\begin{aligned} \left|(Q^n f)(0)\frac{q_n(x-n\alpha)}{x-n\alpha}\right| &= \left|\frac{(Q^n f)(0)}{n}\right| \left|n\frac{q_n(x-n\alpha)}{x-n\alpha}\right| \\ &\le \left|\frac{(Q^n f)(0)}{n}\right| \qquad \text{by the remark at the end of section 3} \\ &\le \frac{\|\Delta^n f\|}{n}. \end{aligned}$$

Since $f$ has period $p$ we can write $f(x)$ in the form $f(x) = \sum \beta_\zeta \zeta^x$ where $\zeta$ runs through the $p$-th roots of unity.
It follows that $(\Delta^n f)(x) = \sum \beta_\zeta (\zeta - 1)^n \zeta^x$, hence $\lim_{n\to\infty} \frac{\|\Delta^n f\|}{n} = 0$ and the uniform convergence of (8) follows.
Taking the limit of (8) for $x \to 0$ and noting that $f'(0) = 0$ we obtain our result which is valid for functions $f$ with period $p$ and $\alpha \in p\mathbf{Z}_p$:

$$\sum_{n=1}^{\infty} \frac{(Q^n f)(0)}{n} q_n(n\alpha) = 0 \tag{9}$$

**EXAMPLE 4.2.** We simply write down a few formulas which are either special cases of (9) or which can be obtained by similar calculations. We leave the details to the reader.

$$\sum_{n=1}^{\infty} \frac{(\Delta^n f)(0)}{n} \binom{n\alpha}{n} = 0 \qquad \alpha \in p\mathbf{Z}_p,\ f \text{ a function with period } p.$$

In $Q_2$: $\quad \sum_{n=1}^{\infty} \frac{(-2)^n}{n} \binom{\alpha n}{n} = 0 \quad$ for $\quad \alpha \in 2\mathbf{Z}_2$

In $Q_2$: $\quad \sum_{n=1}^{\infty} \frac{2^n}{n} \binom{\alpha n}{n} = 0 \quad$ for $\quad \alpha \in 1+2\mathbf{Z}_2$

In $Q_3$: $\sum\limits_{n=0}^{\infty} \frac{(-3)^n}{2n+1}\binom{(2n+1)\alpha}{2n} = -\frac{1}{2}$ for $\alpha \in 2+3\mathbb{Z}_3$.

**EXAMPLE 4.3.** As final example we prove the following formula

$$\frac{\Gamma'_p(x+1)}{\Gamma_p(x+1)} = \Gamma'_p(0) - \sum_{n=1}^{\infty} \frac{u_n}{n}\binom{x+\alpha n}{n} \qquad x \in \mathbb{Z}_p,\ \alpha \in p\mathbb{Z}_p$$

where $\Gamma_p(x)$ is the $p$-adic gamma function (see [3] section 35) and where the integers $u_n$ are defined by the formal power series

$$\frac{(1+t)^{p-1}}{(1+t)^p - t^p} = \sum_{n=0}^{\infty} u_n t^n.$$

Recall that $\Gamma_p(x+1)$ satisfies the following functional equation

$$\begin{aligned} \Gamma_p(x+1) &= -x\Gamma_p(x) && \text{if } |x| = 1 \\ &= -\Gamma_p(x) && \text{if } |x| < 1. \end{aligned}$$

Takling the logarithmic derivative we get

$$\Delta\left\{\frac{\Gamma'_p(x)}{\Gamma_p(x)}\right\} = \frac{\chi(x)}{x} \text{ with } \begin{aligned} \chi(x) &= 1 && \text{if } |x| = 1 \\ &= 0 && \text{if } |x| < 1. \end{aligned}$$

Example 2 of the introduction, with $f(x) = \chi(x)$, gives

$$\Delta\left\{\frac{\Gamma'_p(x)}{\Gamma_p(x)}\right\} = \frac{\chi(x)}{x} = \sum_{n=1}^{\infty} \frac{(\Delta^n\chi)(\alpha n)}{n}\binom{x-1-\alpha n}{n-1}.$$

Taking the indefinite sum and replacing $x$ by $x+1$, we get

$$\frac{\Gamma'_p(x+1)}{\Gamma_p(x+1)} = c + \sum_{n=1}^{\infty} \frac{(\Delta^n\chi)(\alpha n)}{n}\binom{x-\alpha n}{n}.$$

For $x = \alpha = 0$ we find $c = \frac{\Gamma'_p(1)}{\Gamma_p(1)} = -\Gamma'_p(1) = \Gamma'_p(0)$.

Since $\chi$ is a function with period $p$ it is clear that $(\Delta^n\chi)(\alpha n) = (\Delta^n\chi)(0)$ for $\alpha \in p\mathbb{Z}_p$. Henc e

$$\frac{\Gamma'_p(x+1)}{\Gamma_p(x+1)} = \Gamma'_p(0) + \sum_{n=1}^{\infty} \frac{(\Delta^n\chi)(0)}{n}\binom{x-\alpha n}{n}.$$

In order to calculate the numbers $(\Delta^n\chi)(0)$ we use a well-known formula from the calculus of finite differences.

$$\sum_{n=0}^{\infty}\chi(n)u^n = \frac{1}{1-u} - \frac{1}{1-u^p} = \sum_{n=0}^{\infty}(\Delta^n\chi)(0)\frac{u^n}{(1-u)^{n+1}}.$$

Multiplying by $(1-u)$ and taking $\frac{u}{1-u} = t$ we get

$$1 - \frac{(1+t)^{p-1}}{(1+t)^p - t^p} = \sum_{n=0}^{\infty}(\Delta^n\chi)(0)t^n.$$

Replacing $\alpha$ by $(-\alpha)$ and writing $(-u_n)$ for $(\Delta^n\chi)(0)$ we obtain the formula we wanted to prove.

If $\log_p$ denotes the $p$-adic logarithm we have (see [3] p.181)

$$\frac{\Gamma_p'(\frac{1}{2})}{\Gamma_p(\frac{1}{2})} - \Gamma_p'(0) = 2\left(\frac{1}{p} - 1\right)\log_p 2.$$

Putting $x = -\frac{1}{2}$ in our formula we get

$$2\left(1 - \frac{1}{p}\right)\log_p 2 = \sum_{n=1}^{\infty}\frac{u_n}{n}\binom{-\frac{1}{2} - \alpha n}{n} \qquad \alpha \in p\mathbf{Z}_p.$$

## REFERENCES


[1] S. Roman. *The Umbral Calculus.* Academic Press, New York, 1984.

[2] G.C. Rota. *Finite Operator Calculus.* Academic Press, New York, 1975.

[3] W.H. Schikhof. *Ultrametric Calculus.* Cambridge University Press, 1984.

[4] L. van Hamme. *Three generalizations of Mahler's expansion for continuous functions on $\mathbf{Z}_p$.* In *p-adic Analysis. Proceedings, Trento 1989*, edited by F. Baldassarri, S. Bosch and B. Dwork, Springer-Verlag (Lect. Notes 1454), Berlin-Heidelberg-New York (1990), 356–361.



VRIJE UNIVERSITEIT BRUSSEL
FACULTY OF APPLIED SCIENCES
PLEINLAAN 2
B-1050 BRUSSELS, BELGIUM

# DYADIC FRAMES FOR INTERMITTENCY. PERTURBED MODELS

O. Iordache

**Nonarchimedean (n.a.) functional analysis is used here to describe perturbed nearly-linear processes and intermittency characterizing extremely non-linear processes.**

**The method relies upon the existence of more scales of time of the studied processes.**

**The starting point is the replacement of real models by n.a. models in which the time pertains to n.a. fields or rings.**

## INTRODUCTION.

N.a. functional analysis represents an adequate tool in the study of problems involving the concept of hierarchy scaling and mutual conditioning.

Valid constructions in the n.a. frame (the geometries of Veronese and Hilbert, the $p$-adic number theory initiated by Hensel and n.a. functional analysis [1,2,13]) are confronted with an upsurge of interest in recent years in the domain of applications. The interest is motivated by advances in the study of very complex systems as for instance spin glasses, nuclear physics, turbulence, biosystems, computer architecture, combinatorial optimization, classification and taxonomy [3–5,14].

Recent works, trying to apply methods of n.a. functional analysis to systems characterized by a hierarchy of conditioning levels refer to chemistry and biosciences [6,7,17]. This work is a part of such an attempt.

The approach presented here is different from others mainly in that it starts from simple real models and introduces a n.a. structure of time and parameters. Thus an enriched mathematical and physical structure results.

## PHYSICAL CONSIDERATIONS

The existence of more than one characteristic time scale is a well known feature of far from equilibrium complex systems. For instance a variety of distinct time scales is displayed by the processes of relaxation in disordered media, non-ideal mixing, drug-membrane interactions etc. Very complex systems exhibit also intermittent domains of behaviour.

To explain such phenomena it is supposed that due to perturbations or to extremely non-linear interactions a self-organization in a hierarchy of scales of time appears. In fact the n.a. character means precisely that there exists a hierarchical organization.

Some examples of practical interest are fast relaxation processes in lattices, transport by activation over energy barriers, transport through structured interfaces or pores, neural networks etc. All such phenomena have been studied using the concept of n.a. time. The method relies upon the replacement of the real time by elements of n.a. fields or rings. The model equations are then replaced by their n.a. correspondents. Consequently n.a. integration methods should be used.

New formalisms are proposed here to describe either nearly-linear phenomena of relaxation (a perturbation method) or extremely non-linear interactions (a dyadic frame).

## PERTURBED MODELS

### *One-compartmental model*

A simple model governing the relaxation process of the variable $y(t)$ is:

$$\frac{dy}{dt} = -ay \tag{1}$$

Here $a$ is a real constant denoting the rate of transfer or of relaxation.

To account for the existence of scales of time, the real time $t$ is replaced by the expansion:

$$T = t + \varepsilon w_1 t + \cdots + \varepsilon^m w_m t + \cdots \tag{2}$$

where $\varepsilon$ is an arbitrary small positive constant and $w_m$ are real constants. (2) is also written as

$$T = [t, w_1 t, \ldots, w_m t, \ldots].$$

The structure of the time $T$ induces a similar structure of the relaxation function $y(t)$ under consideration. $y(t)$ is replaced by the expansion:

$$Y(T) = y_0(t) + \varepsilon y_1(t) + \cdots + \varepsilon^m y_m(t) + \cdots, \tag{3}$$

also denoted

$$Y(T) = [y_0(t), \ldots, y_m(t), \ldots]$$

where the $y_m(t)$ are real functions.

$T$ and $Y(T)$ are considered as elements of an n.a. ordered field $L$. Some properties of $L$ will be presented in [16].

Let $A, B \in L$, $A = [a_0, \ldots, a_m, \ldots]$, $B = [b_0, \ldots, b_m, \ldots]$. Then $A = B$ if $a_m = b_m$ for all $m$. $A < B$ if $a_0 = b_0, \ldots, a_{j-1} = b_{j-1}$, $a_j < b_j$ for some $j$. Note that $L$ is n.a. Indeed let $A = [0, 1, 0, \ldots]$ and $B = [1, 0, \ldots]$. Then $kA < B$ for any integer $k$.

Define $A + B = [a_0 + b_0, \ldots, a_m + b_m, \ldots]$, $AB = [a_0 b_0, a_0 b_1 + a_1 b_0, \ldots]$, $A/B = [a_0/b_0, (a_1 b_0 - a_0 b_1)/b_0^2, \ldots]$. The only invertibles are those elements $B$ for which $b_0 \neq 0$. Observe also that $A(B + C) = AB + AC$.

For any number $r > 1$, an example of an n.a. valuation on $L$ is defined by:

$$|A| = r^{-m}, \quad m = \min\{n; a_n \neq 0\}.$$

Relative to this valuation $L$ is complete.

In the n.a. frame it is necessary to replace the relaxation model (1) by the n.a. differential equation:

$$\frac{dY}{dT} = -AY. \tag{4}$$

Here $T$ and $Y(T)$ are given by (2) and (3) while $a$ is replaced by the n.a. constant $A = [a, 0, \ldots]$. Solutions of (4) could be obtained using a Dieudonné type integral in the n.a. field $L$ [8].

Denote $T = [t, t, \ldots]$, $T^0 = [t - \Delta t, t - \Delta t, \ldots]$. The solution of the difference equation associated with (4) is:

$$Y(T) = Y(T^0) - AY(T^0)(T - T^0) \tag{5}$$

with $Y(T) = [y_0(t), y_1(t), y_2(t), \ldots]$, $Y(T^0) = [y_0(t-\Delta t), y_1(t-\Delta t), y_2(t-\Delta t), \ldots]$.

Equating the coefficients of equal powers of $\varepsilon$ in (5) produces the system of difference equations:

$$y_0(t)-y_0(t-\Delta t) = -ay_0(t-\Delta t)\Delta t$$
$$y_1(t)-y_1(t-\Delta t) = -a(y_0(t-\Delta t)+y_1(t-\Delta t))\Delta t$$
$$y_2(t)-y_2(t-\Delta t) = -a(y_0(t-\Delta t)+y_1(t-\Delta t)+y_2(t-\Delta t))\Delta t$$
$$\ldots\ldots\ldots$$

Letting $\Delta t \to 0$ one obtains:

$$y_0'(t) = -ay_0(t), \quad y_1'(t) = -a(y_0(t)+y_1(t)), \quad y_2'(t) = -a(y_0(t)+y_1(t)+y_2(t)), \ldots$$

With initial conditions $y_0(0) = y_1(0) = y_2(0) = \ldots = a$, this results in

$$y_m(t) = a\exp(-at)L_m(at), \quad m \geq 0 \tag{6}$$

where $L_m$ is the $m$-th Laguerre polynomial.

A general "computed" solution of (4) is:

$$y_c(t) = a\exp(-at)\sum_m q_m L_m(at) \tag{7}$$

where $q_m$ contains a factor $\varepsilon^m$. A similar solution is valid for general initial conditions and values of $w_m$ in $T$. Note that the choice of $T$ and $T^0$ allows one to consider (6) as a basic solution corresponding to a step $\Delta t$ at any level.

Solutions such as (7) suggest to consider the space of *real* functions of the form $y(t) = \sum_m q_m y_m(t)$ (the vector space $E$ spanned by $y_m(t)$).

The coefficients $q_m$ can be obtained using:

$$q_m = \int_0^\infty y(t)L_m(at)dt \tag{8}$$

On the space of all functions $y(t)$ define the norm:

$$\|y\| = r^{-w(y)}$$

where $r > 1$ and:

$$w(y) = \min\{m; \mu_m(y) \neq 0\}, \quad \mu_m(y) = \int_0^\infty y(t)L_m(at)dt.$$

The coefficients $q_m$ ensure the minimum of the distance between experiments and correlation [7]:

$$d = \|y - \sum_m q_m y_m\|. \tag{9}$$

The distance is null if:

$$\mu_m(y - \sum_m q_m y_m) = 0$$

that is if (8) is verified.

The norm defined on the Banach space $E$ has the following properties:

i. $\|ky\| = \|y\|$ for any real $k$, $\|y+z\| \leq \max\{\|y\|, \|z\|\}$, $\|yz\| \leq \|y\|\,\|z\|$.

ii. $\|y_m(at)\| = r^{-m}$.

iii. $\|\sum_m q_m y_m(at)\| = \max_m \|y_m(at)\|$ for any real $q_m$.

According to i. the valuation is trivial while according to iii. the family $\{y_m\}$ is an orthogonal basis. Consequently any element $y \in E$ can be represented as a convergent series (7).

The topology of $E$ is induced by the metric $d(y, z) = \|y-z\|$. Note that the distance of two functions $y$ and $z$ is less than or equal to $r^{-m}$ iff $\mu_i(y) = \mu_i(z)$ for $i = 0, 1, \ldots, m-1$.

*Discrete models*

A discrete version of the one-compartmental model is presented now. In this case the discrete time $n$ is replaced by the expansion:

$$N = n + \varepsilon w_1 n + \cdots + \varepsilon^m w_m n + \cdots$$

that is by $N = [n, w_1 n, \ldots, w_m n, \ldots]$.

Moreover $y(t)$ is replaced by $Y(N) = [y_0(n), y_1(n), \ldots, y_m(n), \ldots]$ where $y_m(n)$ are real-valued functions.

The discrete version of (1) is:

$$\frac{Y(N) - Y(N^0)}{N - N^0} = -AY(N^0),$$

or equivalently:

$$Y(N) = Y(N^0) - AY(N^0)(N - N^0). \tag{10}$$

Let $N = [n, n, \ldots]$, $N^0 = [n-1, n-1, \ldots]$. Equating the coefficients of different powers of $\varepsilon$ in (10) a system of difference equations results:

$$y_0(n) - y_0(n-1) = -ay_0(n-1)$$

$$y_1(n) - y_1(n-1) = -a(y_0(n-1) + y_1(n-1))$$

$$\ldots\ldots\ldots$$

$$y_m(n) - y_m(n-1) = -a(y_0(n-1) + y_1(n-1) + \cdots + y_m(n-1))$$

$$\ldots\ldots\ldots$$

Denote $\alpha = 1-a$ and consider the initial conditon:

$$y_0(0) = \ldots = y_m(0) = \ldots = 1.$$

One obtains:

$$y_m(n) = \alpha^n M_m(n, 1, \alpha)/m! \tag{11}$$

Here $M_m(n, 1, \alpha)$ are the Meixner polynomials (with $\beta = 1$ in [11]). The Meixner polynomials are orthogonal with the weight $\alpha^n$.

The general "computed" solution is:

$$y_c(n) = \alpha^n \sum_m q_m M_m(n, 1, \alpha)/m! \tag{12}$$

The coefficients $q_m$ result from:

$$(1-\alpha)^{-1}\alpha^{-m}q_m = \sum_n y(n)M_m(n, 1, \alpha)/m!$$

*Stochastic features*

Now in $T$ or $N$ let the $w_m$ be random variables. In this case a model like (12) is replaced by:

$$y(n, \omega) = \alpha^n \sum_m q_m(\omega) M_m(n, 1, \alpha)/m!$$

where $\omega \in \Omega$ and $\Omega$ is a probability space.

The mean of $y(n, \omega)$ is:

$$E(y(n, \omega)) = \alpha^n \sum_m E(q_m(\omega)) M_m(n, 1, \alpha)/m!$$

The variance is:

$$V(y(n,\omega)) = \alpha^{2n} \sum_m V(q_m(\omega))(M_m(n,1,\alpha)/m!)^2.$$

A more difficult problem is to determine the statistics of $y(n)$ at different scales of time of the process. Considering that $\omega$ is random we are faced with an n.a. stochastic equation:

$$Y(N) = Y(N^0) - A(\omega)Y(N^0)(N-N^0).$$

Denote by $r_i$ the probability that the value of $\alpha$ is $\alpha_i$. The mean of $y_0(n)$ is:

$$E(y_0(n)) = \sum_i r_i\alpha_i E(y_0(n-1)) = \overline{\alpha}E(y_0(n-1))$$

where $\overline{\alpha} = \sum_i r_i\alpha_i$. The second order moment is:

$$E(y_0^2(n)) = \sum_i r_i\alpha_i^2 E(y_0^2(n-1)) = \overline{\alpha^2}E(y_0^2(n-1))$$

where $\overline{\alpha^2} = \sum_i r_i\alpha_i^2$.

Using $\overline{\alpha}$ instead of $\alpha$ in the system of difference equations the mean and the variance of $y_m(n)$ could be obtained.

## DYADIC FRAME

### *Model of relaxation*

Suppose that the relaxation described by equation (1) takes place through a hierarchically structured frame. Different levels in this hierarchy are indexed by $m = 0, 1, \ldots$ Making a choice of the unit of time consider that the transport time is $2^m$ for a transition of the type $m$.

On account of the supposed mechanism of transfer through structured media the time $t$ and the rate of transfer $a$ will be considered as elements of a ring $K$. Denote these elements by $T$ and $A$ respectively. Thus: $T = t_0t_1\ldots$, $A = a_0a_1\ldots$ where:

$$T = \sum_j t_j 2^j, \qquad A = \sum_j a_j 2^j \tag{13}$$

where $t_j$ and $a_j$ are the digits "0" or "1". Such expansions give detailed descriptions of the time and of the rates of transfer.

In the new frame the model (1) becomes (4) with $T$ and $A$ given by (13). The formal solution of (4) is, in $K$, given by:

$$Y(T) = Y(T^0) - A \otimes Y(T^0) \otimes (T - T^0). \tag{14}$$

The product $\otimes$ and the addition $\oplus$ in $K$ are defined by:

$$T \otimes A = \left( \sum_j t_j a_j \right) \text{mod } 2, \quad T \oplus A = \sum_j [(t_j + a_j) \text{mod } 2] 2^j. \tag{15}$$

The product rule appears as a natural coupling of the potentialities of transfer $a_j$ and of the available times $t_j$.

The addition rule signifies for instance that, due to physical restrictions, two identical steps could be only forward and backward, and this corresponds to the null time.

Note that we take $T^0 = 000\ldots$ and $Y(T^0) = 100\ldots$ in (14).

Observe that $Y(T)$ given by (14) has only two possible values, corresponding to the null element and the unit element in $K$. The physical significance is failure and succes of the transport process, respectively.

Notice that by simple linear transformations the solutions (14) reduce to the Walsh-type functions defined by:

$$W_m(A, T) = (-1)^{\sum_{j=0}^{m-1} a_j t_j}.$$

Consequently for fixed $A$, the general "computed" solution of (4) is:

$$y_c(T) = \sum_{m=0}^{m-1} q_m W_m(A, T). \tag{16}$$

The Fourier-Walsh coefficients $q_m$ result from:

$$q_m \sum_T W^2_{m(A,T)} = \sum_T y(T) W_m(A, T)$$

where $y(T)$ is the experimentally recorded function [9].

On the space of functions $y(T)$ define the n.a. semi-norm [12, 13]:

$$\|y\| = 2^{-w(y)}$$

where:

$$w(y) = \min\{m; \mu_m(y) \neq 0\}; \quad \mu_m(y) = \sum_T y(T) W_m(A, T).$$

The coefficients $q_m$ ensure the minimum of the distance:

$$d = \|y(T) - \sum_m q_m W_m(A, T)\|.$$

*Stochastic features*

An adequate frame for probability considerations on n.a. spaces is the n.a. measure theory [10, 13, 15].

Consider now $K$ as a fixed set whose elements are dyadic expansions such as (13); for instance $T = t_0 t_1 \ldots$

A collection $\Omega_0$ of subsets of $K$ is a covering ring if for $U_1, U_2 \in \Omega_0$ their intersection, union and difference belong to $\Omega_0$.

Let us define an additive measure $M : \Omega_0 \to F$ where $F$ is another n.a. valued ring. The ring $F$ is defined on $\{0, 1\}$. The elements of $F$ are of the form $V = \sum_j v_j 2^{-j}$, $v_j \in \{0, 1\}$ denoted also by $V = (v_0 v_1 v_2 \ldots)$. The addition on $F$ is componentwise, the rules being $0+1 = 1+0 = 1+1 = 1$; $0+0 = 0$ corresponding to the logical operation OR. The product is the usual multiplication. Some elements of $F$ are $p_0 = (100\ldots)$, $p_1 = (010\ldots)$, $p_2 = (001\ldots)$ and so on.

A valuation on $F$ is given by $|V| = r^{-m}$, $m = \min\{n; v_n \neq 0\}$, $r > 1$.

We take for instance $M(V) = p_k$ where $k$ is the level allowing a transition from a starting state in $K$, for instance $000\ldots$ to the state $U \in K$, in the hierarchical tree naturally associated to $K$.

In the particular case of three levels $k = 1, 2, 3$, considering also 000 to be the starting state, we have $M(001) = p_1$, $M(011) = M(010) = p_2$, $M(100) = M(101) = M(110) = M(111) = p_3$.

Observe that:

$$|p_k + p_j| \leq \max\{|p_k|, |p_j|\}; \quad |p_k p_j| = |p_k|\,|p_j|.$$

Define:

$$\|W\|_M = \max\{|M(U)| : U \in \Omega_0,\ U \subset W\}$$

and

$$N_M(a) = \min\{\|W\|_M : a \in W\}.$$

The index $N_M(a)$ allows a partition of $K$ into disjoint classes.

Probabilities of transition as well as associated entropy could be defined and used for classification and lumping [7].

## BIBLIOGRAPHY

[1] Monna, A.F.: *Analyse Non-Archimédienne*, Springer, Berlin, 1970.

[2] Schikhof, W.H.: *Ultrametric Calculus*, Cambridge Univ. Press, Cambridge, 1984.

[3] Rammal, R., Toulouse, G., Virasoro, M.A.: Rev. Mod. Phys., **58**, (1986), 765–788.

[4] Blumen, A., Klafter, J., Zumofen, G.: in Zschokke I., (ed) *Optical Spectroscopy of Glasses*, Reidel, Dordrecht, (1986), p.199–265.

[5] Freund, P.G.O., Witten, E.: Phys. Lett. B **199**, (1987), 191.

[6] Iordache, O., Isopescu, R., Isopescu, R., Iacobini, A., Jineascu, Ghe.: Chem. Engng. Sci., **43**, (1988), 693–697, (**43**), 1988, 699–703.

[7] Iordache, O.: Rev. Roum. Phys., **33**, (1988), 1159–1162, **34**, (1989), 217–227, Rev. Roum. Biochim., **25**, (1988), 275–281, **25**, (1988), 305–312, **25**, (1988), 325–328, **26**, (1989), 213–222, **26**, (1989), 321–324, **27**, (1990), 43–49.

[8] Dieudonné, J.: Bull. Sc. Math., **68**, (1944), 79–95.

[9] Harmuth, H.F.: *Sequency Theory, Foundation and Applications*, Academic Press, New York, 1977.

[10] Van Rooij, A.C.M., Schikhof, W.H.: Proc. Kon. Ned. Akad. v. Wetensch. A **72**, (1969), 190–199.

[11] Erdélyi, A., Magnus, W., Oberhettinger, F., Tricomi, F.G.: Higher Transcendental Functions, McGraw-Hill, New York, 1953.

[12] Robert, P.: Compositio Math., **19**, (1968), 1–77.

[13] Van Rooij, A.C.M.: *Non-Archimedean Functional Analysis*, Marcel Dekker, New York, 1978.

[14] Volovich, I.V.: Theor. Math. Phys., **71**, (1987), 337–340.

[15] Monna, A.F., Springer, T.A.: Indag. Math., **25**, (1963), 634–642, 643–653.

[16] Lightstone, A.H., Robinson, A.: *Non-Archimedean Fields and Asymptotic Expansions*, North-Holland, Amsterdam, 1975.

[17] Iordache, O.: *Polystochastic Models in Chemical Engineering*, VNU-Science Press, Utrecht, 1987.

DEPARTMENT OF CHEMICAL ENGINEERING
POLYTECHNICAL INSTITUTE
BUCHAREST, ROMANIA

# NON-ARCHIMEDEAN $\Lambda$-NUCLEAR SPACES

A.K. Katsaras

## INTRODUCTION

Throughout this paper, $K$ will be a non-Archimedean, non-trivially valued, complete field. For $p$ a non-Archimedean seminorm on a vector space $E$ over $K$, we will denote by $E_p$ the quotient space $E/\ker p$ and $\pi_p : E \to E_p, x \longmapsto [x]_p$ the canonical surjection. On $E_p$ we will consider the non-Archimedean norm $\|[x]_p\| = p(x)$ and will let $\widehat{E}_p$ denote the completion of $E_p$. For $p, q$ seminorms on $E$, we write $p \prec q$ to mean that there exists $d > 0$ such that $p \leq dq$. For locally convex spaces $E, F$ over $K$, we will denote by $\mathcal{L}(E, F)$ the space of all continuous linear mappings from $E$ to $F$.

As it is well known, if $E, F$ are normed spaces over $K$, then a $T \in \mathcal{L}(E, F)$ is compact iff there exist a sequence $(g_n)$ in $E'$ and a sequence $(y_n)$ in $F$ such that

$$Tx = \sum_{n=1}^{\infty} g_n(x) y_n, \text{ for all } x \in E, \text{ and } \|g_n\| \, \|y_n\| \to 0.$$

This is equivalent to the following: There exist bounded sequences $(h_n)$ and $(y_n)$, in $E'$ and $F$ respectively, and $(\lambda_n) \in c_0$ such that

$$(*) \qquad Tx = \sum_{n=1}^{\infty} \lambda_n h_n(x) y_n.$$

In this paper, we will study operators $T$ which are represented as in $(*)$ where $(\lambda_n)$ belongs to some sequence space $\Lambda$. We will introduce also the $\Lambda$-nuclear spaces and we will study some of their properties.

## 1. PRELIMINARIES

We will recall first the definition of a non-Archimedean Köthe sequence space. By a Köthe set we will mean a set $P$ of sequences $\alpha = (\alpha_n)$, $\alpha_n \geq 0$, such that: i) For each $n$ there exists $\alpha \in P$ with $\alpha_n > 0$. ii) Given $\alpha, \alpha' \in P$ there exists $\beta \in P$ with $\beta_n \geq \alpha_n, \alpha'_n$ for all $n$.

We will write $\alpha \prec \beta$ to mean that there exists $d > 0$ such that $\alpha_n \leq d\beta_n$ for all $n$. For $\alpha \in P$ and $\zeta = (\zeta_n)$ a sequence in $K$, we define $p_\alpha(\zeta) = \sup_n \alpha_n|\zeta_n|$.
The non-Archimedean Köthe sequence space $\Lambda(P)$ is defined by

$$\Lambda(P) = \{\zeta \in K^N : p_\alpha(\zeta) < \infty \text{ for all } \alpha \in P\}.$$

Each $p_\alpha$ is a non-Archimedean seminorm on $\Lambda = \Lambda(P)$ and the space $\Lambda$, equipped with the topology generated by the family $\{p_\alpha : \alpha \in P\}$ is a Hausdorff complete locally convex space. By $\Lambda_0 = \Lambda_0(P)$ we will denote the subspace of $\Lambda$ consisting of all $\zeta$ for which $\alpha_n|\zeta_n| \to 0$ for each $\alpha \in P$. For each $n$, we will let $e^n$ be the sequence with $(e^n)_n = 1$ and $(e^n)_k = 0$ if $k \neq n$. For a sequence $x = (x_n)$ of elements of $K$, we will denote by $|x|$ the sequence $(|x_n|)$.

For $\gamma = (\gamma_n)$ an increasing sequence of non-negative reals and $0 < \rho_0 \leq \infty$, we will denote by $\Lambda_{\gamma,\rho_0}$ the Köthe sequence space corresponding to the Köthe set $P$ consisting of all sequences $(\rho^{\gamma_n})$, $0 < \rho < \rho_0$. For $\rho > 0$, we will denote by $q_\rho$ the seminorm $p_\alpha$ where $\alpha_n = \rho^{\gamma_n}$. The space $\Lambda_{\gamma,\rho_0}$ is metrizable. By [4, Theorem 3.4] $\Lambda = \Lambda_{\gamma,\rho_0}$ is nuclear iff $\gamma$ is not bounded. In this case $\Lambda = \Lambda_0$. The following theorem is from [4].

**1.1. THEOREM.** *Let $\Lambda = \Lambda(P)$ and let $\widetilde{P}$ be the vector subspace of $K^N$ consisting of all $x = (x_n)$ for which there exists $\alpha \in P$ with $|x| \prec \alpha$. Then $\widetilde{P}$ is isomorphic to $\Lambda'_0$ via the isomorphism*

$$x \longmapsto T_x,\ T_x(\zeta) = \langle \zeta, x \rangle = \sum_{n=1}^{\infty} \zeta_n x_n.$$

In view of the preceding Theorem, we will identify $\Lambda'_0$ with $\widetilde{P}$. Each $x$ in $\Lambda$ defines a non-Archimedean seminorm $p_{|x|}$ on $\Lambda'_0$ by $p_{|x|}(y) = \sup_n |x_n y_n|$.

**1.2. PROPOSITION.** *Let $P = \{\alpha^{(m)} : m = 1, 2, \ldots\}$ where $\alpha^{(m)} \leq \alpha^{(m+1)}$. For $\Lambda_0 = \Lambda_0(P)$, we have $\Lambda'_0 = \{y \in \mathbf{N}^N : p_{|x|}(y) < \infty, \forall x \in \Lambda\}$.*

*Proof.* Let $y \in K^N$ be such that $p_{|x|}(y) < \infty$ for all $x \in \Lambda$. We claim that $y \in \widetilde{P}$. Assume that this is not true. Then, there exist positive integers $n_1 < n_2 < \ldots$ such that $|y_{n_k}| > k a_{n_k}^{(k)}$ for all $k$. We choose $x = (\mu_m)$ in $K^N$ as follows: If $m \notin \{n_1, n_2, \ldots\}$, we take $\mu_m = 0$. If $m = n_k$ and $\alpha_{n_k}^{(k)} = 0$, we take $\mu_m \in K$ such that $|\mu_m y_{n_k}| > k$. For $m = n_k$ with $\alpha_{n_k}^{(k)} \neq 0$, let $\mu_m \in K$ be such that $|\lambda^{-1}\mu_m| \leq 1/\alpha_{n_k}^{(k)} \leq |\mu_m|$, where $|\lambda| > 1$. Now $x \in \Lambda$ but $p_{|x|}(y) > k$ for each $k$.

**1.3. Note.** The hypothesis of the preceding Proposition is satisfied in any power sequence space $\Lambda = \Lambda_{\gamma,\rho_0}$. If $\gamma$ is not bounded, then $\Lambda = \Lambda_0$.

**1.4. PROPOSITION.**

1) *A subset $B$ of $\Lambda = \Lambda(P)$ is bounded if there exists $y \in \Lambda$ such that $B \subset B_y = \{x \in \Lambda : |x| \leq |y|\}$.*
2) *If $y \in \Lambda$ and $A_y = B_y \cap \Lambda_0$, then the polar $A_y^0$ of $A_y$ in $\Lambda_0'$ is the set $D = \{x \in \Lambda_0' : p_{|y|}(x) \leq 1\}$.*
3) *The strong topology on $\Lambda_0'$ is generated by the family $\{p_{|y|} : y \in \Lambda\}$.*

*Proof.* 1) Let $B$ be bounded. Then, $d_n = \sup_{x \in B} |x_n| < \infty$. Let $|\lambda| > 1$ and choose $\lambda_n \in K$ with $|\lambda_n| \leq d_n \leq |\lambda\lambda_n|$. Now $y = (\lambda\lambda_n) \in \Lambda$ and $B \subset B_y$. 2) Clearly $D \subset A_y^0$. On the other hand if $x \in A_y^0$, then $x \in D$ since $p_{|y|}(x) = \sup_n |\langle y_n e^n, x\rangle| \leq 1$. 3) It follows from (1) and (2).

**1.5. LEMMA.** *Let $\Lambda = \Lambda_{\gamma,\infty}$ and let $\zeta \in K^N$. Then, $\zeta \in \Lambda_0$ iff $|\zeta_n|^{1/\gamma_n} \to 0$.*

*Proof.* Suppose that $|\zeta_n|^{1/\gamma_n} \not\to 0$. Then, there exist $\varepsilon > 0$ and positive integers $n_1 < n_2 < \ldots$ such that $|\zeta_{n_k}|^{1/\gamma_{n_k}} \geq \varepsilon$. If $\rho = \varepsilon^{-1}$, then $|\zeta_{n_k}|\rho^{\gamma_{n_k}} \geq 1$, for all $k$, and so $\zeta \notin \Lambda_0$. On the other hand, if $|\zeta_n|^{1/\gamma_n} \to 0$, then given $\rho > 0$ and $0 < \varepsilon < 1$, there exists $n_0$ such that $|\zeta_n| \leq \varepsilon^{\gamma_n/\gamma_1} \cdot \rho^{-\gamma_n}$ if $n \geq n_0$. Now, for $n \geq n_0$, we have $|\zeta_n|\rho^{\gamma_n} \leq \varepsilon$ and so $\zeta \in \Lambda_0$.

**1.6. LEMMA.** *Given $x \in \Lambda = \Lambda_{\gamma,\infty}$, there exists $y \in \Lambda$ with $|y| \geq |x|$ and $y_n \neq 0$.*

*Proof.* Choose $\mu_n \in K$, $0 < |\mu_n| < n^{-\gamma_n}$. Then $(\mu_n) \in \Lambda_0$ by Lemma 1.5. Now it suffices to take $y = (y_n)$ where $|y_n| = \max\{|\mu_n|, |x_n|\}$.

**1.7. LEMMA.** *Let $\Lambda = \Lambda_{\gamma,\infty}$ where $\gamma$ is not bounded. The following are equivalent:*

(1) $\sup_n \gamma_{2n}/\gamma_n = d > \infty$.

(2) *If $x = (x_n)$ and $y = (y_n)$ are in $\Lambda$, then the sequence $z = (z_m)$, where $z_{2n-1} = x_n$ and $z_{2n} = y_n$, is an element of $\Lambda$.*

*Proof.* (1) $\Longrightarrow$ (2). For $\rho > 1$, we have $q_\rho(z) \le \max\{q_\sigma(x), q_\sigma(y)\}$, where $\sigma = \rho^d$.
(2) $\Longrightarrow$ (1). Suppose that $d = \infty$ and let $n_1 < n_2 < \dots$ be such that $\gamma_{2n_k}/\gamma_{n_k} > k$ for all $k$. Let $|\lambda| > 1$ and choose $\lambda_k$ such that $|\lambda^{-1}\lambda_k| \le k^{-\gamma_{n_k}} \le |\lambda_k|$. Take $\mu_{n_k} = \lambda_k$ and $\mu_m = 0$ if $m \notin \{n_1, n_2, \dots\}$. Since $|\lambda_k|^{1/\gamma_{n_k}} \le \frac{1}{k} \cdot |\lambda|^{1/\gamma_{n_k}} \to 0$, we have that $(\mu_n) \in \Lambda_0$ but the sequence $\mu_1, \mu_1, \mu_2, \mu_2, \mu_3, \dots$ is not in $\Lambda_0 = \Lambda$ since $|\mu_{n_k}|^{1/\gamma_{2n_k}} \ge k^{-1/k} \to 1$ as $k \to \infty$.

**1.8. Note.** The condition $\sup \gamma_{2n}/\gamma_n < \infty$ is satisfied for instance if $\gamma_n = n^\delta$ $(\delta > 0)$ or $\gamma_n = \log(n+1)$.

**1.9. LEMMA.** *Let $\Lambda = \Lambda_{\gamma,\infty}$ with $\gamma$ not bounded.*

a) *If $(y_n) \in \Lambda$, then every subsequence of $(y_n)$ is also in $\Lambda$.*

b) *Suppose that $\sup_n \gamma_{2n}/\gamma_n = d < \infty$ and let $x = (x_n) \in \Lambda'$. If $(x_{n_k})$ is a subsequence of $(x_n)$ such that $\sup_k n_k/k < \infty$, then $(x_{n_k}) \in \Lambda'$.*

*Proof.* a) It follows from the fact that for $\rho > 1$ we have $\rho^{\gamma_k} \le \rho^{\gamma_{n_k}}$.
b) Let $m$ be a positive integer with $n_k/k \le 2^m$ for all $k$. Then, $\gamma_{n_k}/\gamma_k \le d^m = \theta$ for all $k$. Since $x \in \Lambda'$, there exist $\rho > 1$ and $M > 0$ such that $|x_m| \le M\rho^{\gamma_m}$ for all $m$. Thus, $|x_{n_k}| \le M\rho^{\gamma_{n_k}} \le M(\rho^\theta)^{\gamma_k}$ and so $(x_{n_k}) \in \Lambda'$ by Theorem 1.1.

**1.10. LEMMA.** *Let $\Lambda = \Lambda_{\gamma,\infty}$, with $\gamma$ not bounded, and suppose that $\sup_n \gamma_{2n}/\gamma_n = d < \infty$. For $m$ a positive integer, the function*

$$\varphi_m : \Lambda^m \to \Lambda,\ (y^{(1)}, \dots, y^{(m)}) \to y,$$

where $y_{(k-1)m+r} = y_k^{(r)}$, $r = 1, 2, \dots, m$, is a topological isomorphism.
*Proof.* We first observe that our hypothesis implies that $\varphi_m$ maps $\Lambda^m$ into $\Lambda$ since for $\rho > 1$ we have

$$(*) \qquad q_\rho(y) \le \max_{1 \le r \le m} q_\sigma(y^{(r)}),\ \sigma = \rho^{d^{m_1}},$$

where $m_1$ is an integer with $2^{m_1} \ge m$. Clearly $\varphi_m$ is one-to-one and it follows form $(*)$ that $\varphi_m$ is continuous. Given $z \in \Lambda$ and setting $z^{(k)} = (z_k, z_{m+k}, z_{2m+k}, \dots)$ we have that $z^{(k)} \in \Lambda$ and $\varphi_m(z^{(1)}, \dots, z^{(m)}) = z$. For $\rho > 1$, we have $q_\rho(z^{(k)}) \le q_\rho(z)$. This proves that $\varphi_m$ is onto and that $\varphi_m^{-1}$ is continuous.

In the next Lemma, we will let $\Lambda'_b$ denote the strong dual of $\Lambda$.

**1.11. LEMMA.** *Let $\Lambda = \Lambda_{\gamma,\infty}$ with $\gamma$ not bounded, and suppose that $\sup_n \gamma_{2n}/\gamma_n < \infty$. For $m$ a positive integer, the mapping*

$$\psi_m : (\Lambda'_b)^m \to \Lambda', \ (x^{(1)}, \dots, x^{(m)}) \to x,$$

where $x_{(k-1)m+r} = x_k^{(r)}$ for $r = 1, 2, \dots, m$, is a topological isomorphism.

*Proof.* Let $x = \psi_m(x^{(1)}, \dots, x^{(m)})$ and let $z = (z_n) \in \Lambda$. If $(z^{(1)}, \dots, z^{(m)}) = \varphi_m^{-1}(z)$ , then $p_{|z|}(x) = \max_{1 \le k \le m} p_{|z^{(k)}|}(x^{(k)})$, which proves that $x \in \Lambda'$ (by Proposition 1.2) and that $\psi_m$ is continuous by Proposition 1.4. Given $y = (y_n)$, take $y^{(k)} = (y_k, y_{m+k}, y_{2m+k}, \dots)$. By Lemma 1.9 (b), we have $y^{(k)} \in \Lambda'$. Moreover $y = \psi_m(y^{(1)}, \dots, y^{(m)})$. Finally, $\psi_m^{-1}$ is continuous since, for $z^{(1)}, \dots, z^{(m)} \in \Lambda$, $z = \varphi_m(z^{(1)}, \dots, z^{(m)})$ and $x = \psi_m(x^{(1)}, \dots, x^{(m)})$, we have

$$\max_{1 \le k \le m} p_{|z^{(k)}|}(x^{(k)}) = p_{|z|}(x).$$

We omit the proof of the following easily established.

**1.12. LEMMA.** *Let $\Lambda = \Lambda_{\gamma,\infty}$ with $\gamma$ not bounded. For $x = (x_n) \in \Lambda$ and $y = (y_n) \in \Lambda'_b$ we have $x = \sum_{n=1}^{\infty} x_n e^n$ in $\Lambda$ and $y = \sum_{n=1}^{\infty} y_n e^n$ in $\Lambda'_b$.*

We will finish this section by recalling the definition of the approximation numbers $\alpha_n(T)$, of a linear operator $T$, which were introduced by the author in [3]. For $E, F$ non-Archimedean normed spaces, $T \in \mathcal{L}(E, F)$ and $n$ a nonnegative integer, we define $\alpha_n(T)$ to be the infimum of all numbers $\|T - A\|$ with $A \in \mathcal{L}(E, F)$ with dimension of the range of $A$ less than or equal to $n$.

## 2. Λ-NUCLEAR MAPPINGS

Throughout the rest of the paper, $\Lambda$ will be a non-Archimedean power sequence space $\Lambda_{\gamma,\infty}$ with $\gamma$ not bounded.

**2.1. DEFINITION.** Let $E, F$ be locally $K$-convex spaces and let $T \in \mathcal{L}(E, F)$. The mapping $T$ is called:

1) $\Lambda$-nuclear if there exist an equicontinuous sequence $(g_n)$ in $E'$, a bounded sequence $(y_n)$ in $F$ an element $(\lambda_n)$ of $\Lambda$ such that

$$Tx = \sum_{n=1}^{\infty} \lambda_n g_n(x) y_n, \text{ for all } x \in E.$$

2) $\Lambda$-quasi-nuclear if, for each continuous seminorm $q$ on $F$ there exist $(\lambda_n) \in \Lambda$ and an equicontinuous sequence $(g_n)$ in $E'$ such that

$$q(Tz) \leq \sup_n |\lambda_n g_n(z)|, \text{ for all } z \in E.$$

**2.2. DEFINITION.** Let $T \in \mathcal{L}(E, F)$, where $E, F$ are non-Archimedean normed spaces. Then, $T$ is called of type $\Lambda$ if there exists $(\lambda_n) \in \Lambda$ such that $\alpha_{n-1}(T) \leq |\lambda_n|$ for $n = 1, 2, \ldots$.

**2.3. PROPOSITION.** *Let $T \in \mathcal{L}(E, F)$, where $E, F$ are normed spaces. If $T$ is $\Lambda$-nuclear, then $T$ is of type $\Lambda$.*

*Proof.* There exist $(\lambda_n) \in \Lambda$, $g_n \in E'$, $y_n \in F$, with $\|g_n\| \leq 1$, $\|y_n\| \leq 1$, such that $Tz = \sum_{n=1}^{\infty} \lambda_n g_n(z) y_n$. For each $n$, let $A_n : E \to F$, $A_n(z) = \sum_{k=1}^{n-1} \lambda_k g_k(z) y_k$. Then

$$\alpha_{n-1}(T) \leq \|T - A_n\| \leq \sup_{k \geq n} |\lambda_k|.$$

Let $|\lambda| > 1$ and let $\mu_m \in K$ with $|\lambda|^{-1}|\mu_m| \leq \alpha_{m-1}(T) \leq |\mu_m|$. Now $(\mu_n) \in \Lambda$. In fact, for $\rho > 1$, we have

$$|\mu_m|\rho^{\gamma_m} \leq |\lambda| \sup_{k \geq m} |\lambda_k|\rho^{\gamma_m} \leq |\lambda| \sup_{k \geq m} |\lambda_k|\rho^{\gamma_k} \leq |\lambda| q_\rho(z)$$

where $z = (\lambda_n) \in \Lambda$. This clearly completes the proof.

**2.4. PROPOSITION.** *Let $T \in \mathcal{L}(E, F)$, where $E, F$ are normed spaces over $K$. If*

(*) $$\sup_n \alpha_{n-1}(T) \rho^{\gamma_{n^2}} < \infty$$

*for each $\rho > 0$, then $T$ is $\Lambda$-nuclear.*

*Proof.* Since $\gamma_n \to \infty$, taking $\rho > 1$ in (*) we get that $\alpha_n(T) \to 0$. As it is shown in [5], there exist $d > 0$ and a sequence $(B_n)$ in $\mathcal{L}(E, F)$ such that

$$B_n x = \sum_{k=1}^{2n-1} g_k^{(n)}(x) y_k^{(n)},$$

where $g_k^{(n)} \in E'$, $y_k^{(n)} \in F$ with $\|g_k^{(n)}\| \leq 1$, $\|y_k^{(n)}\| \leq d\alpha_{n-1}(T)$ and $Tx = \sum_{n=1}^{\infty} B_n x$. Setting $h_m = g_k^{(n)}$ and $z_m = y_k^{(n)}$ for $m = (n-1)^2 + k$, $1 \leq k \leq 2n-1$ we have that $Tx = \sum_{m=1}^{\infty} h_m(x) z_m$. To finish the proof, it suffices to show that $\sup_m \|z_m\| \rho^{\gamma_m} < \infty$ for each $\rho > 0$. So, let $\rho > 1$. For $m = (n-1)^2 + k$, $1 \leq k \leq 2n-1$, we have

$$|z_m| \rho^{\gamma_m} \leq d\alpha_{n-1}(T) \rho^{\gamma_{n^2}}$$

and the result follows.

**2.5. COROLLARY.** *Let $T \in \mathcal{L}(E, F)$ be of type $\Lambda$. If $\sup_n \gamma_{n^2}/\gamma_n = d < \infty$, then $T$ is $\Lambda$-nuclear.*

*Proof.* It follows easily from the preceding Proposition.

**2.6. EXAMPLE.** The condition $\sup_n \gamma_{n^2}/\gamma_n < \infty$ holds if $\gamma_n = \log(n+1)$.

Using Lemma 1.7, we get easily the following

**2.7. PROPOSITION.** *Suppose that $\sup \gamma_{2n}/\gamma_n < \infty$. If $T_1, T_2 \in \mathcal{L}(E, F)$ are $\Lambda$-nuclear (resp. $\Lambda$-quasi-nuclear), then $T_1 + T_2$ is $\Lambda$-nuclear (resp. $\Lambda$-quasi-nuclear).*

We have also the following easily established

**2.8. PROPOSITION.** *Let $T_1 \in \mathcal{L}(E, F)$ and $T_2 \in \mathcal{L}(F, G)$. If one of the $T_1, T_2$ is $\Lambda$-nuclear (resp. $\Lambda$-quasi-nuclear), then $T_2 T_1$ is $\Lambda$-nuclear (resp. $\Lambda$-quasi-nuclear).*

**2.9. THEOREM.** *Let $E$ be locally $K$-convex, $F$ a Banach space and $T \in \mathcal{L}(E, F)$. The following are equivalent:*

(1) *$T$ is $\Lambda$-nuclear;*

(2) *$T$ is $\Lambda$-quasi-nuclear;*

(3) *There exist $T_1 \in \mathcal{L}(E, \Lambda)$ $\Lambda$-nuclear and $T_2 \in \mathcal{L}(\Lambda, F)$ such that $T = T_2 T_1$;*

(4) *$T = T_2 T_1$ where $T_1 \in \mathcal{L}(E, \Lambda)$ is $\Lambda$-quasi-nuclear and $T_2 \in \mathcal{L}(\Lambda, F)$.*

*Proof.* Clearly (1) implies (2). (2) $\Longrightarrow$ (3). Let $(\lambda_n) \in \Lambda$ and $(g_n)$ an equicontinuous sequence in $E'$ such that $\|Tz\| \leq \sup_n |\lambda_n g_n(z)|$. Let $|\lambda| > 1$ and choose $\mu_n \in K$ with $|\mu_n| \leq \sqrt{|\lambda_n|} \leq |\lambda \mu_n|$. Since

$$|\mu_n| \rho^{\gamma_n} \leq \sqrt{|\lambda_n| (\rho^2)^{\gamma_n}} \to 0,$$

we have that $(\mu_n) \in \Lambda$. Now define

$$T_1 : E \to \Lambda,\ z \longmapsto (\lambda_n g_n(z)).$$

Taking $\zeta_n = 0$ if $\lambda_n = 0$ and $\zeta_n = \lambda_n \mu_n^{-1}$ if $\lambda_n \neq 0$, we have that $(\zeta_n) \in \Lambda$ since $|\zeta_n| \leq |\lambda^2 \mu_n|$. Now

$$T_1 z = \sum_{n=1}^{\infty} \zeta_n g_n(z) y^{(n)},$$

where $y^{(n)} = \mu_n e^n$, and so $T_1$ is $\Lambda$-nuclear. Let now $\rho > 1$. The normed space $G = (\Lambda, q_\rho)$ is of countable type. If $G_1$ is the closure of $T_1(E)$ in $\widehat{G}$, then (by [8, Theorem 3.16 (v)]) there exists a continuous projection $Q$ of $\widehat{G}$ onto $G_1$. The well defined function $S_1 : T_1(E) \to F$, $S(T_1 z) = Tz$, is continuous since

$$\|Tz\| \leq \sup_n |\lambda_n g_n(z)| \leq \sup_n |\lambda_n g_n(z)| \rho^{\gamma_n} = q_\rho(T_1 z).$$

We will denote the continuous extension of $S_1$ to all of $G_1$ also by $S_1$. Let $T_2 = S_1 Q S_2$ where $S_2 : \Lambda \to \widehat{G}$ is the inclusion map. Clearly $T = T_2 T_1$. Since (3) $\Longrightarrow$ (4) trivially while (3) $\Longrightarrow$ (1) and (4) $\Longrightarrow$ (2) by Proposition 2.8, the result follows.

**2.10. THEOREM.** *Let $E, F$ be locally convex spaces, with $F$ sequentially complete, and let $T \in \mathcal{L}(E, F)$. If $T$ is $\Lambda$-nuclear, then there exist $T_1 \in \mathcal{L}(E, \Lambda'_b)$ $\Lambda$-quasi-nuclear and $T_2 \in \mathcal{L}(\Lambda'_b, F)$ such that $T = T_2 T_1$. The converse is also true if $F$ is a Banach space.*

*Proof.* Let $y = (\lambda_n) \in \Lambda$, $(g_n)$ an equicontinuous sequence in $E'$ and $(y_n)$ a bounded sequence in $F$ such that $Tz = \sum_{n=1}^{\infty} \lambda_n g_n(z) y_n$. For each $z \in E$, $(g_n(z))$ is an element of $\Lambda'_b$ and so we get a map $T_1 : E \to \Lambda'_b$, $T_1(z) = (g_n(z))$. Let $q$ be a continuous seminorm on $E$ with $|g_n| \leq q$. For $x \in \Lambda$, taking $h_n = x_n g_n$, we have $\|h_n\|_q \leq |x_n|$ and $p_{|x|}(T_1 z) = \sup |h_n(z)|$, which shows that $T_1$ is $\Lambda$-quasi-nuclear. For each $\zeta = (\zeta_n) \in \Lambda'_b$, the series $\sum_{n=1}^{\infty} \lambda_n \zeta_n y_n$ is convergent in $F$ since $\lambda_n \zeta_n y_n \to 0$ and $F$ is sequentially complete. Define $T_2 : \Lambda'_b \to F$, $T_2 \zeta = \sum_{n=1}^{\infty} \lambda_n \zeta_n y_n$. Since for each continuous seminorm $p$ on $F$, we have $p(T_2 \zeta) \leq [\sup_n p(y_n)] p_{|y|}(\zeta)$, $T_2$ is continuous. Moreover $T = T_2 T_1$. The last assertion follows from Proposition 2.8 and Theorem 2.9.

Let $\Lambda[E']$ denote the family of all sequences $(g_n)$ in $E'$ for which there exist a continuous seminorm $p$ on $E$ and $(\lambda_n) \in \Lambda$ such that $\|g_n\|_p \leq |\lambda_n|$. Clearly every

$(g_n) \in \Lambda[E']$ is equicontinuous. For $\omega = (g_n) \in \Lambda[E']$, we define $p_\omega$ on $E$ by $p_\omega(z) = \sup_n |g_n(z)|$. Each such $p_\omega$ is a continuous seminorm on $E$. If $\sup_n \gamma_{2n}/\gamma_n < \infty$, then given $\omega_1, \omega_2 \in \Lambda[E']$ there exists $\omega \in \Lambda[E']$ such that $p_\omega \geq p_{\omega_1}, p_{\omega_2}$.

**2.11. THEOREM.** *Let $T \in \mathcal{L}(E,F)$ where $E,F$ are locally convex spaces. If $T$ is $\Lambda$-quasi-nuclear, then $T$ is continuous with respect to the topology $\tau$ of uniform convergence on the members of $\Lambda[E']$. The converse is also true if $\sup_n \gamma_{2n}/\gamma_n < \infty$.*

*Proof.* Since $\tau$ is generated by the family $\mathcal{F} = \{p_\omega : \omega \in \Lambda[E']\}$ of seminorms on $E$, the first part follows directly from the definitions. In case $\sup \gamma_{2n}/\gamma_n < \infty$, the converse is also true since in this case $\mathcal{F}$ is directed.

## 3. $\Lambda$-NUCLEAR SPACES

**DEFINITION.** A locally convex space $E$ is called $\Lambda$-nuclear if, for each continuous seminorm $p$ on $E$, there exists another one $q$, with $p \prec q$, such that the canonical mapping $\varphi_{pq} : \widehat{E}_q \to \widehat{E}_p$ is $\Lambda$-nuclear.

**3.1. PROPOSITION.** *For each locally convex space $E$, the space $(E, \sigma(E,E'))$ is $\Lambda$-nuclear.*

*Proof.* It follows easily from the fact that for each weakly continuous seminorm $p$ on $E$ the normed space $E_p$ is finite dimensional.

**3.2. THEOREM.** *A locally convex space $E$ is $\Lambda$-nulcear iff, for each continuous seminorm $p$ on $E$, there exists $\omega \in \Lambda[E']$ such that $p \leq p_\omega$.*

*Proof.* ($\Longrightarrow$). Given $p$, let $q$ be as in the definition of $\Lambda$-nuclearity. Since $\varphi_{pq}$ is $\Lambda$-nuclear, there exists $(\lambda_n) \in \Lambda$ and $(g_n) \subset (\widehat{E}_q)'$, with $\|g_n\| \leq |\lambda_n|$, such that

$$\|\varphi_{pq}(z)\| \leq \sup_n |g_n(z)| \qquad (z \in \widehat{E}_q).$$

Let $h_n \in E'$, $h_n = g_n \circ \pi_q$. For $x \in E$, we have $|h_n(x)| \leq |\lambda_n| q(x)$ and so $\|h_n\|_q \leq |\lambda_n|$. Moreover, $p(z) = \|[z]_p\| \leq \sup |h_n(z)|$. Thus $p \leq p_\omega$, where $\omega = (h_n) \in \Lambda[E']$.

($\Longleftarrow$). Given $p$ continuous seminorm on $E$, let $\omega = (h_n) \in \Lambda[E']$ with $p \leq p_\omega$. Let $(\lambda_n) \in \Lambda$ and $q$ a continuous seminorm on $E$ be such that $\|h_n\|_q \leq |\lambda_n|$. Since $(\lambda_n)$ is bounded, we have that $p \prec q$. Let $g_n : E_q \to K$, $g_n([z]_q) = h_n(z)$. Then $g_n$

is well defined and $\|g_n\| \leq |\lambda_n|$. For $z \in E$, we have $\|[z]_p\| = p(z) \leq \sup |g_n([z]_q)|$. Denoting the extension of $g_n$ to all of $\widehat{E}_q$ also by $g_n$, we have $\|\varphi_{pq}(w)\| \leq \sup_n |g_n(w)|$ for all $w \in \widehat{E}_q$ and so $\varphi_{pq}$ is $\Lambda$-nuclear by Theorem 2.9.

We omit the proof of the following easily established

**3.3. PROPOSITION.** *If $E$ is $\Lambda$-nuclear, then $E$ is nuclear and every subspace of $E$ is $\Lambda$-nuclear.*

**3.4. THEOREM.** *For a locally convex space $E$, consider the following statements:*

(1) *$E$ is $\Lambda$-nuclear.*

(2) *For every locally convex space $F$, every $T \in \mathcal{L}(E, F)$ is $\Lambda$-quasi-nuclear.*

(3) *For every Banach space $F$, every $T \in \mathcal{L}(E, F)$ is $\Lambda$-nuclear.*

(4) *The topology of $E$ coincides with the topology of uniform convergence on the members of $\Lambda[E']$.*

*Then,* (1) $\Longleftrightarrow$ (2) $\Longleftrightarrow$ (3) $\Longrightarrow$ (4). *If* $\sup \gamma_{2n}/\gamma_n < \infty$, *then* (4) $\Longleftrightarrow$ (1).

*Proof.* (1) $\Longrightarrow$ (2). Let $q$ be a continuous seminorm on $F$. There exists a continuous seminorm $p$ on $E$ such that $q(Tz) \leq p(z)$. Since $E$ is $\Lambda$-nuclear, there exists $(g_n) = \omega \in \Lambda[E']$ with $p \leq p_\omega$. It follows from this that $T$ is $\Lambda$-quasi-nuclear.

(2) $\Longrightarrow$ (3) and (1) $\Longrightarrow$ (4) by the Theorems 2.9 and 3.2 respectively. In case $\sup \gamma_{2n}/\gamma_n < \infty$, the family $\{p_\omega : \omega \in \Lambda[E']\}$ is directed and the implication (4) $\Longrightarrow$ (1) follows from Theorem 3.2. Finally, assume (3) and let $p$ be a continuous seminorm on $E$. Since the mapping $\pi_p : E \to \widehat{E}_p$ is $\Lambda$-nuclear, there exists $(g_n) \in \Lambda[E']$ such that $p(z) = \|[z]_p\| \leq \sup |g_n(z)|$, and so $E$ is $\Lambda$-nuclear by Theorem 3.2.

**3.5. PROPOSITION.** *If $H$ is a closed subspace of a $\Lambda$-nuclear space $E$, then the quotient space $E/H$ is $\Lambda$-nuclear.*

*Proof.* Let $T \in \mathcal{L}(E/H, F)$ where $F$ is a Banach space. If $\pi : E \to E/H$ is the quotient map, then $S_1 = T \circ \pi$ is $\Lambda$-nuclear and hence there exist $(\lambda_n) \in \Lambda$, $p$ a continuous seminorm on $E$ and $g_n \in E'$ such that $\|g_n\|_p \leq |\lambda_n|$ and $\|S_1 x\| \leq \sup_n |g_n(x)|$. We may assume that $\|S_1 x\| \leq p(x)$ for all $x \in E$. If $S_2 : E_p \to F$ is the induced map and if $\widetilde{g}_n$ is defined on $E_p$ by $\widetilde{g}_n([x]_p) = g_n(x)$, then $\|\widetilde{g}_n\| \leq |\lambda_n|$ and $|S_2([x]_p) \leq \sup_n |\widetilde{g}_n([x]_p)|$. Let $\widehat{\widetilde{g}}_n$ and $\widehat{S}_2$ be the extension of $\widetilde{g}_n$ and $S_2$ on all of $\widehat{E}_p$. Since $E$ is $\Lambda$-nuclear, the space $\widehat{E}_p$ is of countable type. Let $|\lambda| > 1$. By [8, Theorem 3.16 (V)], there exists a projection $Q$ of $\widehat{E}_p$ onto the closure $G$ of $\pi_p(H)$ in $\widehat{E}_p$, with $\|Q\| \le |\lambda|$. For $w \in E_p$ we have $\|S_2(w)\| \le \sup |\widetilde{g}_n(w)|$ and therefore $\|S_2(w)\| \le \sup |\widehat{\widetilde{g}}_n(w)|$ for $w \in \widehat{E}_p$. Let now $z \in E$. We have

$$\begin{aligned}\|T(\pi(z))\| = \|\widehat{S}_2([z]_p)\| &= \|\widehat{S}_2([z]_p - Q([z]_p)\| \\ &\le \sup |\widehat{\widetilde{g}}_n([z]_p - Q([z]_p))|.\end{aligned}$$

Putting $\varphi_n(w) = \widehat{\widetilde{g}}_n(w - Qw)$, for $w \in \widehat{E}_p$, we have that $\|T(\pi(z))\| \le \sup |\varphi_n([z]_p)|$. Let $h_n(z) = \varphi_n([z]_p)$ and let $\widetilde{h}_n$ be defined on $E/H$ by $\widetilde{h}_n(\pi(z)) = h_n(z)$. We have that $\|T(\pi(z))\| \le \sup |\widetilde{h}_n(\pi(z))|$. Moreover, for $w = [z]_p - Q([z]_p)$,

$$|\widetilde{h}_n(\pi(z))| = |\widehat{\widetilde{g}}_n(w)| \le |\lambda_n| \, \|w\| \le |\lambda_n \lambda| p(z).$$

Thus $|\widetilde{h}_n(\pi(z))| \le |\lambda_n \lambda| \widetilde{p}(\pi(z))$ where $\widetilde{p}(\pi(z)) = \inf\{p(x) : x - z \in H\}$. This proves that $T$ is Λ-quasi-nuclear. The result now follows from Theorems 2.9 and 3.4.

**3.6. THEOREM.** *Let the sequence $\gamma = (\gamma_n)$ be such that $\sup \gamma_{2n}/\gamma_n < \infty$. If $\{E_i : i \in I\}$ is a family of Λ-nuclear spaces, then the topological product $E = \Pi E_i$ is Λ-nuclear.*

*Proof.* Let $p$ be a continuous seminorm on $E$. There are $i_1, \ldots, i_m$ in $I$ and for each $k$ a continuous $p_k$ on $E_{i_k}$ such that $p(x) \le \max_{1 \le k \le m} p_k(x_{i_k})$ for $x \in E$. For each $k$, there exist $(\lambda_{kn}) \in \Lambda$, $(g_{kn}) \subset E'_{i_k}$ and a continuous seminorm $q_k$ on $E_{i_k}$ such that $\|g_{kn}\|_{q_k} \le |\lambda_{kn}|$ and $p_k(x_{i_k}) \le \sup_n |g_{kn}(x_{i_k})|$ for $x_{i_k} \in E_{i_k}$. For $j = (n-1)m + k$, $1 \le k \le m$, take $\lambda_j = \lambda_{kn}$ and $h_j : E \to K$, $h_j(x) = g_{kn}(x_{i_k})$. If the seminorm $q$ on $E$ is defined by $q(x) = \max_{1 \le k \le m} q_k(x_{i_k})$, then $q$ is continuous and $\|h_j\|_q \le |\lambda_j|$. Moreover, $p(x) \le \sup_j |h_j(x)|$. Since $(\lambda_j) \in \Lambda$ (by Lemma 1.10), the result follows from Theorem 3.2.

**3.7. PROPOSITION.** *Let $E$ be a Hausdorff locally convex space and consider the following statements:*

(1) *For every Banach space $F$ and every $T \in \mathcal{L}(E, F)$ there exist $T_1 \in \mathcal{L}(E, \Lambda)$ and $T_2 \in \mathcal{L}(\Lambda, F)$ such that $T = T_2 T_1$.*

(2) *If $\mathcal{P} = \{p_i : i \in I\}$ is a generating family of continuous seminorms on $E$, then $E$ is topologically isomorphic to a subspace of the cartesian product $\Lambda^I$.*

(3) *$E$ is topologically isomorphic to a subspace of some cartesian product $\Lambda^J$.*

*Then, (1) $\Longrightarrow$ (2) $\Longrightarrow$ (3). If $\sup \gamma_{2n}/\gamma_n < \infty$, then (3) $\Longrightarrow$ (1).*

*Proof.* (1) $\Longrightarrow$ (2). Take $G_i = \widehat{E}_{p_i}$ and let $\varphi_i : E \to G_i$ be the quotient map. By (1), $\varphi_i = S_i T_i$ with $T_i \in \mathcal{L}(E, \Lambda)$ and $S_i \in \mathcal{L}(\Lambda, G_i)$. The mapping

$$\varphi : E \to G = \Lambda^I, \ \varphi(x) = (T_i x)_{i \in I},$$

is one-to-one since $E$ is Hausdorff. Let $q_i$ be a continuous seminorm on $\Lambda$ such that $\|S_i \zeta\| \leq q_i(\zeta)$. Given $i \in I$, the seminorm $q(x) = q_i(x_i)$ is continuous on $G$ and, for $z \in E$, $p_i(z) = \|[z]_{p_i}\| \leq q_i(T_i z) = q(\varphi(z))$, which proves that $\varphi^{-1}$ is continuous.

Suppose now that (3) holds and that $\sup \gamma_{2n}/\gamma_n < \infty$. We may assume that $G$ is a subspace of $\Lambda^J$. Let $T \in \mathcal{L}(E, F)$, where $F$ is a Banach space. There are $i_1, \ldots, i_m$ in $J$ and $\rho > 1$ such that

$$\|Tz\| \leq \max_{1 \leq k \leq m} \sup_n \left| z_n^{(i_k)} \right| \rho^{\gamma_n} \qquad (z = (z^{(i)}) \in E.$$

The function

$$h : E \to \Lambda^m, \ h(z) = \left( z^{(i_1)}, \ldots, z^{(i_m)} \right),$$

is continuous. Let $T_1 = \varphi_m h$ where $\varphi_m$ is defined as in Lemma 1.10. Let $M = (\Lambda, q_\rho)$ and let $\pi : \Lambda \to \widehat{M}$ be the inclusion map. The function $f : \pi(T_1(E)) \to F$, $f(\pi(T_1 z)) = Tz$ is well defined. Since $\rho > 1$ and $(\gamma_n)$ is increasing, it is clear that

$$q_\rho(T_1 z) \geq \max_{1 \leq k \leq m} q_\rho \left( z^{(i_k)} \right) \geq \|Tz\|,$$

and so $f$ is continuous. Let $H$ be the closure of $\pi(T_1(E))$ in $\widehat{M}$ and let $\widehat{f}$ be the continuous extension of $f$ to all of $H$. Since $\widehat{M}$ is of countable type, there exists a continuous projection $Q$ of $\widehat{M}$ onto $H$. Taking $T_2 = \widehat{f} Q \pi$, we have $T = T_2 T_1$.

**3.8. THEOREM.** *Let $E$ be a Hausdorff locally convex space and consider the following statements:*

(1) *$E$ is $\Lambda$-nuclear.*

(2) a) *If $\{p_i : i \in I\}$ is a generating family of continuous seminorms on $E$, then $E$ is topologically isomorphic to a subspace of $\Lambda^I$.*

b) *Every $T \in \mathcal{L}(E, \Lambda)$ is $\Lambda$-quasi-nuclear.*

(3) i) *$E$ is topologically isomorphic to a subspace of some product $\Lambda^J$.*

ii) *Every $T \in \mathcal{L}(E, \Lambda)$ is $\Lambda$-quasi-nuclear.*

*Then,* (1) $\Longrightarrow$ (2) $\Longrightarrow$ (3). *If* $\sup \gamma_{2n}/\gamma_n < \infty$, *then* (3) $\Longrightarrow$ (1).

*Proof.* (1) $\Longrightarrow$ (2). Let $T \in \mathcal{L}(E, F)$ where $F$ is a Banach space. By Theorems 3.4 and 2.9, $T = T_2T_1$ with $T_1 \in \mathcal{L}(E, \Lambda)$ and $T_2 \in \mathcal{L}(\Lambda, F)$. Now (a) follows from Proposition 3.7 while (b) follows from Theorem 3.4.

Assume next that (3) holds and that $\sup \gamma_{2n}/\gamma_n < \infty$. Let $S \in \mathcal{L}(E, F)$ where $F$ is a Banach space. By Proposition 3.7, there are $S_1 \in \mathcal{L}(E, \Lambda)$ and $S_2 \in \mathcal{L}(\Lambda, F)$ with $S = S_2S_1$. By (ii), $S_1$ is $\Lambda$-quasi-nuclear and so $S$ is $\Lambda$-nuclear by Proposition 2.8 and Theorem 2.9. The result now follows from Theorem 3.4.

**3.9. THEOREM.** *If the Hausdorff locally convex space $E$ is $\Lambda$-nuclear and if $\mathcal{F} = \{p_i : i \in I\}$ is a generating family of continuous seminorms on $E$, then $E$ is topologically isomorphic to a subspace of the cartesian product $(\Lambda'_b)^I$.*

*Proof.* For each $i \in I$, there exist a continuous seminorm $q_i$ on $E$, $p_i \prec q_i$, $(\lambda_n^{(i)}) \in \Lambda$ and $g_n^{(i)} \in E'$, with $\|g_n^{(i)}\|_{q_i} \le |\lambda_n^{(i)}|$ such that $p_i(z) \le \sup_n |g_n^{(i)}(z)|$. Set $h_n^{(i)} = 0$ if $\lambda_n^{(i)} = 0$ and $h_n^{(i)} = [\lambda_n^{(i)}]^{-1} g_n^{(i)}$ is $\lambda_n^{(i)} \neq 0$. Since, for $\zeta \in \Lambda$, $\sup_n |\zeta_n h_n^{(i)}(z)| \le (\sup_n |\zeta_n|) q_i(z)$, the mapping

$$T_i : E \to \Lambda'_b,\ T_i(z) = (h_n^{(i)}(z)),$$

is continuous. Thus, the map $T : E \to (\Lambda'_b)^I$, $Tz = (T_iz)_{i\in I}$, is continuous. This map is also one-to-one since $E$ is Hausdorff. Finally $T^{-1} : T(E) \to E$ is continuous. In fact let $i \in I$ and set

$$G_i = \{\zeta \in \Lambda' : \sup_n |\zeta_n \lambda_n^{(i)}| \le 1\}.$$

If $\pi_i : (\Lambda'_b)^I \to \Lambda'_b$ is the ith projection map, then $T^{-1}(\pi_i^{-1}(G_i)) \subset \{z \in E : p_i(z) \le 1\}$ which completes the proof.

## 4. ANOTHER CHARACTERIZATION OF THE Λ-NUCLEAR SPACES

**4.1. DEFINITION.** Let $E, F$ be locally convex spaces. A continuous bilinear form $B$ on $E \times F$ is called:

a) $\Lambda$-nuclear if there exist equicontinuous sequences $(g_n)$, $(h_n)$ in $E'$ and $F'$, respectively, and $(\lambda_n) \in \Lambda$ such that

$$B(x, y) = \sum_{n=1}^{\infty} \lambda_n g_n(x) h_n(y) \qquad (x \in E,\ y \in F).$$

b) $\Lambda$-quasi-nuclear if there exist $(g_n) \subset E'$, $(h_n) \subset F'$, $(\lambda_n) \in \Lambda$ and continuous seminorms $p, q$ on $E$, $F$, respectively, such that $\|g_n\|_p \|h_n\|_q \leq |\lambda_n|$ and

$$|B(x,y)| \leq \sup_n |g_n(x)h_n(y)|.$$

**4.2. THEOREM.** *For $E$ locally convex, the following are equivalent:*

(1) *$E$ is $\Lambda$-nuclear.*

(2) *$E$ is polar and for every locally convex space $F$, every continuous bilinear form on $E \times F$ is $\Lambda$-nuclear.*

(3) *$E$ is polar and for every locally convex space $F$, every continuous bilinear form on $E \times F$ is $\Lambda$-quasi-nuclear.*

*Proof.* (1) $\Longrightarrow$ (2). Since every $\Lambda$-nuclear space is nuclear and hence strongly polar, $E$ is polar. Let $B$ be a continuous bilinear form on $E \times F$. Let $p, q$ be continuous seminorms on $E, F$, respectively, such that $|B(x,y)| \leq p(x)q(y)$. Since $E$ is $\Lambda$-nuclear, there exists a continuous seminorm $p'$ on $E$, $p \leq p'$, such that the canonical mapping $\varphi = \varphi_{pp'} : \widehat{E}_{p'} \to \widehat{E}_p$ is $\Lambda$-nuclear. Let $(\lambda_n) \in \Lambda$, $g_n \in E'$, $x_n \in \widehat{E}_p$ be such that $|g_n| \leq p'$, $\|x_n\| \leq 1$ and $\varphi([z]_{p'}) = \sum_{n=1}^{\infty} \lambda_n g_n(z) x_n$. The mapping

$$B_1 : E_p \times F \to K, \ B_1([x]_p, y) = B(x,y),$$

is well defined and bilinear. Moreover

$$|B_1([x]_p, y)| \leq \|[x]_p\| q(y).$$

Let $\widehat{B}_1 : \widehat{E}_p \times F \to K$ be the continuous extension of $B_1$. For all $z \in \widehat{E}_p$, $y \in F$, we have $|\widehat{B}_1(z,y)| \leq \|z\| q(y)$. Since, for $x \in E$, $[x]_p = \sum_{n=1}^{\infty} \lambda_n g_n(x) x_n$, we have

$$B(x,y) = \widehat{B}_1([x]_p, y) = \sum_{n=1}^{\infty} \lambda_n g_n(x) \widehat{B}_1(x_n, y).$$

Taking $h_n : F \to K$, $h_n(y) = \widehat{B}_1(x_n, y)$, we have

$$B(x,y) = \sum_{n=1}^{\infty} \lambda_n g_n(x) h_n(y) \quad \text{and} \quad |h_n(y)| \leq q(y).$$

(3) $\Longrightarrow$ (1). Let $p$ be a continuous polar seminorm on $E$ and set $G = (E,p)'$. On $G$ we consider the norm $f \longmapsto \|f\|_p$. The bilinear form $B(x,f) = f(x)$, on

$E \times G$, is continuous and so it is Λ-quasi-nuclear. Let $(\lambda_n) \in \Lambda$, $g_n \in E'$, $\varphi_n \in G'$ and $p'$ a continuous seminorm on $E$ such that $\|\varphi_n\| \, \|g_n\|_{p'} \leq |\lambda_n|$ and $|f(x)| \leq \sup_n |g_n(x)\varphi_n(f)|$. For $f \in E$ with $|f| \leq p$, we have $|\varphi_n(f)| \leq \|\varphi_n\|$. Since $p$ is polar, we get that $p(x) \leq \sup_n |g_n(x)| \, \|\varphi_n\| \leq \sup_n |\lambda\mu_n g_n(x)|$, where $|\lambda| > 1$ and $|\mu_n| \leq \|\varphi_n\| \leq |\lambda\mu_n|$. Taking $h_n = \lambda\mu_n g_n$, we have $\|h_n\|_{p'} \leq |\lambda\lambda_n|$ and $p(x) \leq \sup_n |h_n(x)|$. Now the result follows from Theorem 3.2.

## 5. Λ-NUCLEARITY OF KÖTHE SEQUENCE SPACES

**5.1. PROPOSITION.** *Let $I$ be an index set and let $\zeta = (\zeta_i) \in c_0(I)$. Let $T$ be the map $T_\zeta : c_0(I) \longrightarrow c_0(I)$, $(Tx)_i = \zeta_i x_i$. Then, for each non-negative integer $n$, we have*

$$\alpha_n(T) = \sup_{J \in \mathcal{F}_{n+1}} \inf_{i \in J} |\zeta_i|,$$

*where $\mathcal{F}_{n+1}$ is the collection of all subsets of $I$ containing $n+1$ elements.*

*Proof.* The proof is the same as the one of Proposition 2.9 in [3].

**5.2. THEOREM.** *For a non-Archimedean Köthe sequence space $G = \Lambda(P)$, the following are equivalent: (1) $G$ is Λ-nuclear, (2) $F = \Lambda_0(P)$ is Λ-nuclear, (3) For each $\alpha \in P$, for which $N_\alpha = \{n : \alpha_n \neq 0\}$ is infinite, there exist $\beta \in P$, with $\alpha \prec \beta$, an one-to-one onto function $\pi : N \to N_\alpha$ and $(\lambda_n) \in \Lambda$ such that $\alpha_{\pi(n)} \leq |\lambda_n|\beta_{\pi(n)}$ for all $n$.*

*Proof.* (1) $\Longrightarrow$ (2) by Proposition 3.3. (2) $\Longrightarrow$ (3). Let $\alpha \in P$ with $N_\alpha$ infinite. Since $F$ is Λ-nuclear, there exists $\beta \in P$, with $\alpha \prec \beta$, such that the canonical mapping $\varphi : \widehat{F}_{p_\beta} \to \widehat{F}_{p_\alpha}$ is Λ-nuclear. Since $\Lambda \subset c_0$, $\varphi$ is nuclear and hence (by [4, Theorem 2.2]) there exists a sequence $(\delta_n)$, with $\delta_n \to 0$, such that $\alpha_n \leq \delta_n\beta_n$. Let $|\lambda| > 1$ and let $\mu_n$, $\sigma_n$ in $K$ be such that $|\mu_n| \leq \alpha_n \leq |\lambda\mu_n|$ and $|\sigma_n| \leq \beta_n \leq |\lambda\sigma_n|$. The map

$$\psi_\beta : F_{p_\beta} \to c_0(N_\beta), \ [x]_{p_\beta} \to (\sigma_n x_n)_{n \in N_\beta},$$

is well defined. Moreover, $|\lambda|^{-1}\|[x]_{p_\beta}\| \leq \|\psi_\beta[x]_{p_\beta}\| \leq \|[x]_{p_\beta}\|$ and so $\psi_\beta$ is continuous and one-to-one. Let $\widehat{\psi}_\beta$ denote the continuous extension of $\psi_\beta$ to all of $\widehat{F}_{p_\beta}$. Since $\|\widehat{\psi}_\beta(z)\| \geq |\lambda|^{-1}\|z\|$ for $z \in \widehat{F}_{p_\beta}$, $\widehat{\psi}_\beta$ is one -to-one and $\widehat{\psi}_\beta^{-1} : M = \widehat{\psi}_\beta(\widehat{F}_{p_\beta}) \to \widehat{F}_{p_\beta}$ is continuous. It follows that $M$ is complete and hence $M = c_0(N_\beta)$ since $M$ contains $\psi_\beta(F_{p_\beta})$ which is dense in $c_0(N_\beta)$. Thus $\widehat{\psi}_\beta$ is an onto homeomorphism. Next we take

$$\psi_\alpha : F_{p_\alpha} \to c_0(N_\beta),\ [x]_{p_\alpha} \to (\mu_n x_n)_{n\in \mathbf{N}_\beta}.$$

This mapping is well defined and continuous. Let $\widehat{\psi}_\alpha$ be the continuous extension of $\psi_\alpha$ to all of $\widehat{F}_{p_\alpha}$ and take $S : c_0(N_\beta) \to c_0(N_\beta)$, $S = \widehat{\psi}_\alpha \varphi \widehat{\psi}_\beta^{-1}$. Since $|\mu_n \sigma_n^{-1}| \le |\lambda|\delta_n \to 0$ and $N_\alpha$ is infinite, there exists $\pi : N \to N_\alpha$ one-to-one, onto, such that $|\tau_{\pi(n)}| \ge |\tau_{\pi(n+1)}|$, where $\tau_k = \sigma_k^{-1}\mu_k$. By Proposition 2.8, $S$ is $\Lambda$-nuclear and hence of type $\Lambda$ by Proposition 2.3. Let

$$S_1 : c_0(N_\beta) \to c_0(N_\beta),\ y \longmapsto (\sigma_n^{-1}\mu_n y_n)_{n\in \mathbf{N}_\beta}.$$

Since $S_1$ is continuous and $S(y) = S_1(y)$ for $y = \psi_\beta([x]_{p_\beta})$ we have $S = S_1$ since $\psi_\beta(F_{p_\beta})$ is dense in $c_0(N_\beta)$. In view of Proposition 5.1, we have $\alpha_m(S) = \alpha_m(S_1) = |\tau_{\pi(m+1)}|$. If $\lambda_k = \lambda\tau_{\pi(k)}$, then $(\lambda_k) \in \Lambda$ since $S$ of type $\Lambda$. Moreover $\alpha_{\pi(n)} \le |\lambda_n|\beta_{\pi(n)}$.

(3) $\Longrightarrow$ (1). Let $\alpha \in P$. There are two possible cases:

**Case I.** $N_\alpha = \{n_1, n_1, \ldots, n_m\}$. Let $|\lambda| > 1$ and let $\lambda_k \in K$, $|\lambda_k| \le \alpha_{n_k} \le |\lambda\lambda_k|$, $k = 1, \ldots, m$. Take $g_{n_k} \in G'$, $g_{n_k}(\zeta) = \lambda\lambda_k\zeta_{n_k}$. Now $p_\alpha(\zeta) \le \max\limits_{1\le k\le m} |g_{n_k}(\zeta)|$.

**Case II.** $N_\alpha$ is infinite. Let $\beta \in P$, $(\lambda_n) \in \Lambda$ and $\pi : N \to N_a$ be as in (3). For each $n$, let $\mu_n \in K$, $|\mu_n| \le \beta_{\pi(n)} \le |\lambda\mu_n|$. Take $g_n \in G'$, $g_n(\zeta) = \lambda\lambda_n\mu_n\zeta_{\pi(n)}$. Then $\|g_n\|_{p_\beta} \le |\lambda\lambda_n|$ and $p_\alpha(\zeta) \le \sup\limits_n |g_n(\zeta)|$. Now the result follows from Theorem 3.2.

**5.3. PROPOSITION.** *Suppose that for each $\alpha \in P$ there exist $\beta \in P$, $(\lambda_n) \in \Lambda$ and an one-to-one mapping $j : N \to N$ such that $\alpha_n \le \beta_n|\lambda_{j(n)}|$ for all $n$. Then $G = \Lambda(P)$ is $\Lambda$-nuclear.*

*Proof.* Let $|\lambda| > 1$ and $|\mu_n| \le \beta_n \le |\lambda\mu_n|$. For $k = j(n)$, take $g_k \in G'$, $g_k(\zeta) = \lambda\mu_n\lambda_k\zeta_n$, while for $k \notin j(N)$ we take $g_k = 0$. Now $\|g_k\|_{p_\beta} \le |\lambda\lambda_k|$ and $p_\alpha(\zeta) \le \sup\limits_n |g_k(\zeta)|$.

**5.4. PROPOSITION.** *Let $\delta = (\delta_n)$ be an increasing sequence of positive numbers with $\delta_n \to \infty$. Then $G = \Lambda_{\delta,\infty}$ is $\Lambda_{\gamma,\infty}$-nuclear iff there exist $\rho > 1$, $(\lambda_n) \in \Lambda$ and a permutation $\pi : N \to N$ such that $|\lambda_n| \ge \rho^{-\delta_{\pi(n)}}$ for all $n$.*

*Proof.* ($\Longrightarrow$). Let $\sigma > 1$. By Theorem 5.2, there exist $(\lambda_n) \in \Lambda$, $\rho > 1$ and a permutation $\pi : N \to N$ such that $\sigma^{\delta_{\pi(n)}} \le |\lambda_n|\rho^{\delta_{\pi(n)}}$. Now $\rho^{-\delta_{\pi(n)}} \le (\sigma/\rho)^{\delta_{\pi(n)}} \le |\lambda_n|$.

(⟸). Let $\sigma > 0$ and take $\rho_1 = \max\{\rho, \sigma\}$. We have

$$(\sigma\rho_1^{-2})^{\delta_{\pi(n)}} \leq (\sigma\rho_1^{-1})^{\delta_{\pi(n)}} \cdot \rho_1^{-\delta_{\pi(n)}} \leq |\lambda_n|$$

and the result follows from Theorem 5.2.

**5.5. COROLLARY.** *If $\delta = (\delta_n)$, with $0 < \delta_n \uparrow \infty$ is such that $\gamma_n/\delta_n \to 0$, then $\Lambda_{\delta,\infty}$ is $\Lambda_{\gamma,\infty}$-nuclear.*

*Proof.* Let $|\lambda| > 1$ and let $\lambda_n \in K$, $|\lambda^{-1}\lambda_n| \leq 2^{-\delta_n} \leq |\lambda_n|$. For $n$ large enough, we have

$$|\lambda_n|^{1/\gamma_n} \leq |\lambda|^{1/\gamma_n} \cdot 2^{-\delta_n/\gamma_n} \leq 2 \cdot 2^{-\delta_n/\gamma_n} \to 0$$

and so $(\lambda_n) \in \Lambda$ by Lemma 1.5. Now the result follows from Proposition 5.4.

**5.6. EXAMPLES.** 1) Let $\gamma_n = \log(n+1)$. Then $\zeta = (\zeta_n) \in \Lambda$ iff the sequence $(|\zeta_n|)$ is rapidly decreasing, i.e. $\sup_n (n+1)^k|\zeta_n| < \infty$ for $k = 0, 1, 2, \ldots$ If $G = \Lambda_{\delta,\infty}$, $\delta_n = n^\alpha$, where $\alpha > 0$ is constant, then $G$ is $\Lambda_{\gamma,\infty}$-nuclear since $\gamma_n/\delta_n \to 0$.

2) If $\sup_n \gamma_{2n}/\gamma_n = d < \infty$, then $\Lambda$ is not $\Lambda$-nuclear. In fact, suppose that $\Lambda$ is $\Lambda$-nuclear. By Proposition 5.4, there exists $\rho > 1$, $(\lambda_n) \in \Lambda$ and a permutation $\pi : N \to N$ such that $\rho^{-\gamma_{\pi(n)}} \leq |\lambda_n|$. By Lemma 1.5, we have $\left[\rho^{-\gamma_{\pi(n)}}\right]^{1/\gamma_n} \to 0$ and so $\gamma_{\pi(n)}/\gamma_n \to \infty$. If $m$ is such that $\gamma_{\pi(n)}/\gamma_n > d$ for $n > m$, then for $n > m$ we have $\pi(n) > 2n > 2m$. Since $\pi$ is onto, we have that $\{1, 2, \ldots, 2m\} \subset \{\pi(1), \ldots, \pi(m)\}$, a contradiction. Thus $\Lambda$ is not $\Lambda$-nuclear. In particular $\Lambda$ is not $\Lambda$-nuclear if $\gamma_n = n^\alpha$ $(\alpha > 0)$ or if $\gamma_n = \log(n+1)$.

Motivated from Theorem 5.2, we give the following

**5.7. DEFINITION.** A non-Archimedean Köthe sequence space $G = \Lambda(P)$ is called universally $\Lambda$-nuclear if there exist $\alpha_0 \in P$ with $N_{\alpha_0} = N$ and a permutation $\pi : N \to N$ such that for all $\alpha \in P$, with $\alpha_0 \prec \alpha$, there is $\beta \in P$, $\alpha \prec \beta$, and $(\lambda_n) \in \Lambda$ such that $\alpha_{\pi(n)} \leq |\lambda_n|\beta_{\pi(n)}$ for all $n$.

Note that if $N_{\alpha_0} = N$, then putting $P_0 = \{\alpha : \alpha_0 \prec \alpha\}$ we have $\Lambda(P) = \Lambda(P_0)$.

**5.8. THEOREM.** *Let $G = \Lambda(P)$ be universally $\Lambda$-nuclear. Then, the strong dual of $G$ is $\Lambda_{\delta,\infty}$-nuclear for each $\delta = (\delta_n)$, $0 < \delta_n \uparrow \infty$, such that $\gamma_n/\delta_n \to \infty$.*

*Proof.* We may assume that $N_\alpha = N$ for all $\alpha \in P$. Let $\pi : N \to N$ be as in the definition of the universal $\Lambda$-nuclearity. Since $\pi$ is one-to-one and onto, we have $\pi(n) \to \infty$ as $n \to \infty$. Since $G$ is $\Lambda$-nuclear (by Theorem 5.2), we have $\Lambda(P) = \Lambda_0(P)$. Let $|\lambda| > 1$ and let $\mu_n \in K$ with $|\lambda^{-1}\mu_n| \leq 2^{\gamma_n} \leq |\mu_n|$. Since $|\mu_n|^{-1/\delta_n} \leq 2^{-\gamma_n/\delta_n} \to 0$, we have that $(\mu_n^{-1}) \in \Lambda_{\delta,\infty}$. Let $x \in \Lambda(P)$. Given $\alpha \in P$, let $\beta \in P$ and $(\lambda_n) \in \Lambda$ be such that $\alpha_{\pi(n)} \leq |\lambda_n|\beta_{\pi(n)}$. We have that $\sup_n |\lambda_n\mu_n| \leq |\lambda| \sup_n |\lambda_n| 2^{\gamma_n} = d < \infty$ and

$$|x_n\mu_{\pi^{-1}(n)}|\alpha_n \leq |x_n\mu_{\pi^{-1}(n)}\lambda_{\pi^{-1}(n)}|\beta_n \leq d|x_n|\beta_n,$$

which proves that $z = (x_n\mu_{\pi^{-1}(n)}) \in \Lambda(P)$. For each $m = \pi^{-1}(n)$, let $g_m : g'_b \to K$, $g_m(\zeta) = z_n\zeta_n\mu_m^{-1}$. We have that $|g_m(\zeta| \leq |\mu_m^{-1}|p_{|z|}(\zeta)$ and $p_{|z|}(\zeta) = \sup_n |g_m(\zeta)|$. Now the result follows from Proposition 1.4 and Theorem 3.2.

**5.9. THEOREM.** *The strong dual $\Lambda'_b$ of $\Lambda = \Lambda_{\gamma,\infty}$ is $\Lambda$-nuclear.*

*Proof.* Let $y = (y_n) \in \Lambda$ and $|\lambda| > 1$. Choose $\lambda_n \in K$, $|\lambda_n| \leq \sqrt{|y_n|} \leq |\lambda\lambda_n|$. Then $z = (\lambda_n) \in \Lambda$ and $|y_n| \leq |\lambda^2\lambda_n|\,|\lambda_n|$. Now the result follows from Propositions 1.2 and 5.3.

**5.10. THEOREM.** *Suppose that* $\sup_n \gamma_{2n}/\gamma_n < \infty$. *Then, for a Hausdorff locally convex space $E$, the following are equivalent:*

(1) *$E$ is $\Lambda$-nuclear;*

(2) *If $\{p_i : i \in I\}$ is a generating family of continuous seminorms on $E$, then $E$ is topologically isomorphic to a subspace of the product $[\Lambda'_b]^I$.*

(3) *$E$ is topologically isomorphic to a subspace of some cartesian product $(\Lambda'_b)^J$.*

*Proof.* By Theorem 5.9, $\Lambda'_b$ is $\Lambda$-nuclear. Our assumption on $\gamma$ implies that $(\Lambda'_b)^J$ is $\Lambda$-nuclear by Theorem 3.6. Now the result follows from Theorem 3.9 and Proposition 3.3.

## REFERENCES


[1] N. De Grande-De Kimpe. *Non-Archimedean Fréchet spaces generalizing spaces of analytic functions.* Nederl. Akad. Wetensch. Indag. Math. **44** (1982), 432–439.

[2] N. De Grande-De Kimpe. *Non-Archimedean topologies of countable type and associated operators.* Nederl. Akad. Wetensch. Indag. Math. **49** (1987), 15–28.

[3] A.K.Katsaras. *On non-Archimedean locally convex spaces.* Bull. Soc. Math. Grèce (N.S.) **29** (1988), 61–83.

[4] A.K. Katsaras. *On non-Archimedean sequence spaces.* Bull. Inst. Math. Acad. Sinica, **18** (2) (1990).

[5] A.K. Katsaras and J. Martínez-Maurica. *Approximation numbers of continuous linear mappings and compact operators on non-Archimedean spaces* (preprint).

[6] G. Pérez-García. *On compactoidity in non-Archimedean locally convex spaces with a Schauder basis.* Nederl. Akad. Wetensch. Indag. Math. **50** (1988), 85–88.

[7] A. Pietsch. *Nuclear locally convex spaces.* Springer-Verlag, 1972.

[8] A.C.M. van Rooij. *Non-Archimedean Functional Analysis.* Marcel Dekker, New York, 1978.

[9] W.H. Schikhof. *Locally convex spaces over non-spherically complete valued fields I, II.* Bull. Soc. Math. Belg. Sér. B **38** (1986), 187–224.

DEPARTMENT OF MATHEMATICS
UNIVERSITY OF IOANNINA
GREECE

# THE LOCALLY $K$-CONVEX SPACES $C^n(X)$, $C^\infty(X)$

Samuel Navarro*

ABSTRACT. The $C^n$-functions have been systematically studied by W. Schikhof [4]. The locally $K$-convex space $C^n(X)$ has been studied principally by N. De Grande-De Kimpe [1]. In this paper we present results characterizing the compactoids in $C^n(X)$, in the spirit of Martínez-Maurica and Navarro [3].

We also study the structure of $C^n(X)$ presenting conditions for its normability and metrizability.

## 1. PRELIMINARIES

Throughout this paper $K$ will denote a complete field with a non trivial, nonarchimedean valuation and $X$ is a nonempty subset of $K$ without isolated points. In the definition of $C^n$ we take the one given by W. Schikhof [4].
For $n \in N$, we set

$$\nabla^n X = \{(x_1, x_2, \ldots, x_n) \in X^n / \text{ if } i \neq j, \text{ then } x_i \neq x_j\}.$$

Also, for a function $f : X \longrightarrow K$ we define, for each $n$ the following quotients:

$$\Phi_n f(x_1, x_2, \ldots, x_n, x_{n+1}) = \frac{\Phi_{n-1} f(x_1, x_3, x_4, \ldots, x_{n+1}) - \Phi_{n-1}(x_2, x_3, \ldots, x_{n+1})}{x_1 - x_2}$$

where $\Phi_0 f = f$.
A function $f : X \longrightarrow K$ is a $C^n$-function ($f \in C^n(X)$), if $\Phi_n f$ can be extended to a continuous function $\overline{\Phi}_n f$ defined on $X^{n+1}$ with values in $K$.

We denote by $C^n(X, K)$ (or simply $C^n$) the family of $C^n$-functions, and by $C^\infty(X, K)$ (or simply $C^\infty$) the intersection of all $C^n$, for $n \in N$.

* Partially supported by Fondecyt-Chile 890744 and DICYT-USACH.

Since the set $X$ is without isolated points, the extension of $\Phi_n f$, for a $C^n$-function, is unique.

$C^n(X)$ are vector spaces closed under products and they satisfy the following property:

$$C(X) \supseteq C^1(X) \supseteq C^2(X) \supseteq \ldots \supseteq C^\infty(X).$$

In $C^n(X)$ we consider the topology defined by the seminorms

$$\|f\|_{B,n} = \max_{i \in \{1,2,\ldots,n\}} \|\Phi_i f\|_B,$$

where $B$ is a compact subset of $X$ and

$$\|\Phi_i f\|_B = \sup_{x \in B^{i+1}} |\overline{\Phi}_i f(x)|.$$

## 2. NORMABILITY OF $C^n(X)$

In [1] N. De Grande-De Kimpe proved that $C^\infty$ is a nuclear space; therefore it is not normable. For the case of $C^n$ we can prove the following result.

**THEOREM 1.** *$C^n(X)$ is normable if and only if $X$ is a compact space.*

*Proof.* Let $X$ be a non compact subset of $K$; let $p$ be a continuous seminorm on $C^n(X)$; we prove that $\ker p \neq \{0\}$. To this end we may suppose that $p = \|\ \ \|_{B,n}$ where $B \subset X$ is compact. Now choose an $x \in X \backslash B$ and a clopen $V$ with $x \in V$ but $V \cap B = \emptyset$. Then the characteristic function of $V$, $\chi_V \in \ker p$. Therefore $X$ must be a compact set. The converse is clear.

## 3. METRIZABILITY OF $C^n(X)$

Recall that a topological space $X$ is hemicompact if it has a fundamental sequence (i.e. cofinal sequence) of compact sets.

In the case of the space of continuous functions, endowed by the compact-open topology, Katsaras [2] proved that $C(X)$ is metrizable if and only if $X$ is hemicompact. We extended this result to $C^n(X)$.

**THEOREM 2.** *$C^n(X)$ is metrizable if and only if $X$ is hemicompact.*

*Proof.* Clearly the condition is sufficient since the fundamental sequence of compact sets determines a sequence of seminorms which define the topology in $C^n(X)$.
For the converse, we consider the countable basis of zero neighborhoods $\{V_{j,m}\}_{j,m\in N}$, where

$$V_{j,m} = \{f \in C^n(X) / \sup_{x\in S_j^{m+1}} |\overline{\Phi}_m f(x)| < \varepsilon_j\};$$

here $\{S_j\}_{j\in N}$ is a sequence of compact sets in $X$, and $\{\varepsilon_j\}$ a real sequence in $]0,1[$. We now let $T$ be a compact subset in $X$; then there exist $j, m$ such that

$$V_{j,m} \subseteq \{f \in C^n(X) / \sup_{x\in T} |f(x)| < \frac{1}{2}\}$$

and thus $T$ is contained in $S_j$. If not, let $x$ in $T - S_j$ and $U$ be a clopen which contains $S_j$ but does not contain $x$. Thus, the function $f = \mathcal{X}_{U^c}$ is in $V_{j,m}$, however $\sup_{x\in T} |f(x)| > \frac{1}{2}$.

## 4. COMPACTOIDS IN $C^n(X)$

Since $C^\infty(X)$ is a nuclear space, then $\mathcal{F} \subseteq C^\infty$ is compactoid if and only if $\mathcal{F}$ is bounded. In the space $C^n(X)$ the situation is different. We first state the following.

**LEMMA 3.** *Let $(Y_i)_{i\in I}$ be a family of locally convex Hausdorff spaces, and let $Y$ be a vector space with the projective topology relative to $\{(Y_i, f_i)\}_{i\in I}$, where $f_i : Y \to Y_i$ is linear for every $i$. If $(f_i)_{i\in I}$ separates points in $Y$, then*

$$F : Y \to \prod_{i\in I} Y_i \text{ where } F(y) = (f_i(y)) \text{ for } y \in Y;$$

*is an isomorphism on $F(Y)$.*

The proof is easy, and we may obtain:

**COROLLARY 4.** *Let $A$ be a subset of $Y$. Then $A$ is compactoid in $Y$ if and only if $f_i(A)$ is compactoid in $Y_i$ for every $i$ in $I$* (here we use the notations of the Lemma).

**COROLLARY 5.** *Let $X$ be a compact subset of $K$ without isolated points, and $\mathcal{F}$ be a subset of $C^n(X)$. Then $\mathcal{F}$ is compactoid if and only if $\overline{\Phi}_k(\mathcal{F})$ is pointwise bounded and equicontinuous in $C(X^{k+1})$, for $k = 0, 1, 2, \ldots, n$.*

*Proof.* From the preliminaries, we know that $C^n(X)$ has the projective topology relative to $C(X^{k+1}, \varphi_k)$ where $\varphi_k(f) = \overline{\Phi}_k(f)$. Then $\mathcal{F}$ is compactoid if and only if for every $k \in \{0, \ldots, n\}$ $\overline{\Phi}_k(\mathcal{F})$ is compactoid in $C(X^{k+1})$. In [3] we proved that $\overline{\Phi}_k(\mathcal{F})$ is compactoid if and only if it is pointwise bounded and equicontinuous.

**PROPOSITION 6.** *Let $X$ be a compact subset of $K$ without isolated points, and $\mathcal{F}$ be in $C^n(X)$. If $\overline{\Phi}_k(\mathcal{F})$ is bounded, then $\overline{\Phi}_{k-1}(\mathcal{F})$ is equicontinuous in $C(X^k)$, for $k = 1, 2, \ldots, n$.*

*Proof.* If $x = (x_1, \ldots, x_k)$ and $y = (y_1, \ldots, y_k)$ are in $X^k$ and $f \in \mathcal{F}$, we have

$$\overline{\Phi}_{k-1}f(x_1, \ldots, x_k) - \overline{\Phi}_{k-1}(y_1, \ldots, y_k) = \sum_{j=1}^{k} (x_j - y_j)(\overline{\Phi}_k f)(x_1, \ldots, x_j, y_j, \ldots, y_k)$$

then

$$\begin{aligned} |\overline{\Phi}_{k-1}f(x_1, \ldots, x_k) - \overline{\Phi}_{k-1}(y_1, \ldots, y_k)| &\leq \|\overline{\Phi}_k f\|_\infty \max_{j=1,\ldots,n} |x_j - y_j| \\ &\leq \|\overline{\Phi}_k f\|_\infty \cdot \|x - y\| \end{aligned}$$

where $\|x - y\| = \max_{j=1,\ldots,n} |x_j - y_j|$. Thus $\overline{\Phi}_{k-1}(\mathcal{F})$ is equicontinuous.

**PROPOSITION 7.** *Let $X$ be a compact subset of $K$ without isolated points, and $\mathcal{F}$ in $C^n(X)$ be pointwise bounded. If $\overline{\Phi}_k(\mathcal{F})$ is equicontinuous, then $\overline{\Phi}_k(\mathcal{F})$ is pointwise bounded in $C(X^{k+1})$.*

*Proof.* If $x = (x_1, \ldots, x_{k+1})$ is in $\nabla^{k+1}(X)$, then by [4] (Ex 29.A) we have

$$\Phi_k f(x) = \Phi_k f(x_1, \ldots, x_{k+1}) = \sum_{i=1}^{k+1} \{\prod_{i \neq j} (x_i - x_j)^{-1}\} f(x_i).$$

This yields

$$\begin{aligned} |\Phi_k f(x)| &\leq \max_{i \in \{1,\ldots,k+1\}} (|f(x_i)| \cdot \prod_{i \neq j} |x_i - x_j|^{-1}) \\ &\leq \max_{i \in \{1,\ldots,k+1\}} (\prod_{i \neq j} |x_i - x_j|^{-1}) \cdot \max_{i \in \{1,\ldots,k+1\}} |f(x_i)|. \end{aligned}$$

We now take $y$ in $X^{k+1} - \nabla^{k+1}X$. Since $\overline{\Phi}_k(\mathcal{F})$ is equicontinuous, if $\varepsilon = 1$ for the corresponding $\delta(\varepsilon)$, we may take $x$ in the dense subset $\nabla^{k+1}X$ such that $\|x-y\| < \delta(\varepsilon)$. Thus we obtain

$$|\overline{\Phi}_k f(y)| = |\overline{\Phi}_k f(y) - \Phi_k f(x) + \Phi_k f(x)| \leq \max\{|\overline{\Phi}_k f(y) - \Phi_k f(x)|, |\Phi_k f(x)|\}.$$

Hence, if $M_x \geq |\Phi_k f(x)|$, for every $f$ in $\mathcal{F}$, then $|\overline{\Phi}_k f(y)| \leq \max\{1, M_x\}$, for every $f$ in $\mathcal{F}$.

**THEOREM 8.** *Let $X$ be a compact subset of $K$ without isolated points, and $\mathcal{F}$ be in $C^n(X)$. Then $\mathcal{F}$ is a compactoid if and only if $\mathcal{F}$ is pointwise bounded and $\overline{\Phi}_n(\mathcal{F})$ is equicontinuous in $C(X^{n+1})$.*

*Proof.* Since $\mathcal{F}$ is pointwise bounded and $\overline{\Phi}_n(\mathcal{F})$ is equicontinuous, then Prop. 7 implies $\overline{\Phi}_n(\mathcal{F})$ pointwise bounded. Thus $\overline{\Phi}_n(\mathcal{F})$ is compactoid. Now, Prop. 6 implies $\overline{\Phi}_{n-1}(\mathcal{F})$ is equicontinuous. Using the above argument we have that $\overline{\Phi}_{n-1}(\mathcal{F})$ is compactoid. Since this holds for all $k$ in $\{0, 1, \ldots, n\}$,we conclude that $\overline{\Phi}_k(\mathcal{F})$ is compactoid, for all $k$ in $\{0, 1, \ldots, n\}$, and by Corollary 4, $\mathcal{F}$ is compactoid in $C^n(X)$. Conversely, if $\mathcal{F}$ is compactoid in $C^n(X)$, then $\overline{\Phi}_k(\mathcal{F})$ is compactoid in $C(X^{k+1})$, for all $k$ in $\{0, 1, \ldots, n\}$. By [3] $\overline{\Phi}_k(\mathcal{F})$ is pointwise bounded, for all $k$ in $\{0, 1, \ldots, n\}$; in particular for zero. Thus $\mathcal{F}$ is pointwise bounded. We also have $\overline{\Phi}_n(\mathcal{F})$ is equicontinuous.

**Remark 9.** If $X$ is not compact, in $C^n(X)$ we consider the topology determined by the seminorms $\| \quad \|_B$, where $B$ is a compact subset of $X$. We also note that $C^n(X)$ has the projective topology related to the applications $\varphi_k$, where $\varphi_k(f) := \overline{\Phi}_k(f)$. We may thus extend Theorem 8 to every $X$ in $K$ without ioslated points.

**ACKNOWLEDGMENT:** The author wishes to thank Professor Javier Martínez-Maurica for his relevant contributions in preparing this work and to the referee for his valuable comments.

## REFERENCES

[1] De Grande-De Kimpe, N.: *The Non-archimedean space $C^\infty(X)$.* Compositio Math. **48** (1983), 297–309.

[2] Katsaras, A.: *The strict topology in non-archimedean vector-valued function spaces*. Nederl. Akad. Wetensch. Indag. Math. **46** (1984), 189–201.

[3] Martínez-Maurica, J., Navarro Hernández, S.: *p-adic Ascoli theorems*. Rev. Mat. Univ. Complut. Madrid **3** (1990), 19–27.

[4] Schikhof W.H.: *Ultrametric calculus*. Cambridge Univ. Press 1984.

DEPARTAMENTO DE MATEMÁTICA
FAC. CIENCIA
U. DE SANTIAGO DE CHILE
CASILLA 5659 C-2,
SANTIAGO, CHILE

# THE HAHN-BANACH EXTENSION PROPERTY IN $p$-ADIC ANALYSIS

C. Pérez-García*

**ABSTRACT.** In 1952 A.W. Ingleton proved that every locally convex space over a spherically complete valued field has the Hahn-Banach extension property. This paper is a survey of the more important results obtained by different authors in the last years about the extension property for locally convex spaces over non-spherically complete fields and for non-locally convex spaces. Also, we include some new contributions of the author on this subject.

## 0. PRELIMINARIES

Throughout this paper $K$ is a non-archimedean non-trivially valued complete field with valuation $|.|$.

Let $X$ be a $K$-vector space. Recall that a *non-archimedean seminorm* on $X$ is a map $p : X \longrightarrow [0, +\infty)$ with the properties:

i) $p(\alpha x) = |\alpha| p(x) \quad (\alpha \in K,\ x \in X)$.

ii) $p(x+y) \leq \max(p(x), p(y)) \quad (x, y \in X)$.

According to [25] $p$ is called *polar* if

$$p = \sup\{|f| : f \in X^*,\ |f| \leq p\}$$

(where $X^*$ denotes the algebraic dual of $X$).

In the definition of a non-archimedean seminorm the strong triangle inequality (property ii)) is a natural but by no means logically unavoidable requirement. Let us

* Research partially supported by DGICYT (grant PS87-0094)

say that a function $p: X \to [0, +\infty)$ is a *seminorm* on $X$ if it has propery i) while in addition to that one has

$$p(x+y) \leq p(x) + p(y) \qquad (x, y \in X)$$

A set $A \subset X$ is called *absolutely convex* if $\alpha x + \beta y \in A$ whenever $x, y \in A$, $\alpha, \beta \in K$, $|\alpha|, |\beta| \leq 1$. For $A \subset X$, $c_0(A)$ will denote its absolutely convex hull and $[A]$ its linear hull. $A$ is *edged* if it is absolutely convex and for each $x \in X$ the set $\{|\lambda| : \lambda x \in A\}$ is closed in $|K|$. Following [1] a set $A \subset X$ is said to be *quasiconvex* if $a_1 A + \ldots + a_n A \subset A$ for every $a_1, \ldots, a_n \in K$ with $|a_1| + \ldots + |a_n| \leq 1$ and for every $n \in N$.

Let $U \subset X$ be a quasiconvex and absorbent subset of $X$. For each $n \in N$ we define $q_n : X \to [0, +\infty)$ by

$$q_n(x) = \inf \left\{ \sum_{i=1}^{n} |a_i| : a_i \in K, \ x \in \sum_{i=1}^{n} a_i U \right\} \qquad (x \in X).$$

Then, $q_U : X \to [0, +\infty)$ given by

$$q_U(x) = \inf\{q_n(x) : n \in N\} \qquad (x \in X)$$

is a seminorm on $X$ such that (see [1])

$$\{x \in X : q_U(x) < 1\} \subset U \subset \{x \in X : q_U(x) \leq 1\}$$

and

$$q_U \leq q_1 \leq \rho . q_U$$

where $\rho = \inf\{|\lambda| : \lambda \in K, \ |\lambda| > 1\}$. It is called the *calibre* of $U$. Observe that if the valuation on $K$ is dense and $U$ is absolutely convex, then $q_U = q_1$ where $q_1$ is the Minkowski functional of $U$ according to [19], p.31.

A topological vector space $X$ over $K$ is said to be *locally quasiconvex* if its topology is defined by a family of seminorms (equivalently, if it has a fundamental system consisting of quasiconvex zero neighbourhoods). By $X'$ we denote the topological dual of $X$.

**EXAMPLES**

**1.** Locally quasiconvex spaces over $K$ whose topology is defined by a family of non-archimedean seminorms. This kind of spaces will be called ***non-archimedean locally convex spaces.*** They are the traditional locally convex spaces considered in the non-archimedean literature. But in a natural way we can think of some spaces which are not included in this class:

**2.** ([32]) Let $\ell^q$ be the vector space of all $K$-valued sequences $(a_1, a_2, \ldots)$ for which $\|(a_n)\|_q = (\sum |a_i|^q)^{1/q} < \infty$ $(q \geq 1)$.

**3.** ([30]) More in general, let $\Phi : [0, +\infty) \to [0, +\infty)$ be an Orlicz function (e.g. $\Phi(t) = t^q$, $q \geq 1$) and let $\ell^\Phi$ be the vector space of all $K$-valued sequences $(a_1, a_2, \ldots)$ for which

$$\|(a_n)\|_\Phi = \inf\{r \in (0, +\infty) : \sum \Phi(|a_i|/r) \leq 1\} < \infty.$$

**4.** ([32]) Let $L^q([0,1], K)$ be the vector space of all Borel measurable functions $f : [0,1] \to K$ for which the Lebesgue integral $\|f\|_q = (\int_0^1 |f(x)|^q dx)^{1/q}$ is finite $(q \geq 1)$.

**5.** ([6]) Let $(\Omega, \sum, \mu)$ be a measure space, $(E, \|.\|)$ be a non-archimedean Banach space over $K$ and $\Phi : [0, +\infty) \to [0, +\infty)$ be an Orlicz function. We consider the vector space $L^\Phi(\Omega, \mu, E)$ of all measurable functions for which

$$\|f\|_\Phi = \inf\{s > 0 : \int_\Omega \Phi(\|f\|/s) d\mu \leq s\} < \infty.$$

All of these last examples are particular cases of the more general class consisting of all normed spaces over $K$ whose norm does not verify the strong triangle inequality. According to [32] this kind of spaces will be called ***A-normed spaces.***

Although $A$-normed spaces appear in a natural way, their properties were very little known for a long time. In 1974 *A.F. Monna* [20] proposed the general theory of this kind of spaces as an open problem in non-archimedean analysis.

Recently *J. Martínez-Maurica* and the author studied various topics (some of them referred to the extension of continuous linear functionals) within the context of $A$-normed spaces. The results obtained in relation to this subject are included in references [10]-[16] and [22] of the bibliography. They contain improvements of some of the results previously proved by *A.F. Monna* in [17] and [18].

## 1. WEAKLY CLOSED SUBSPACES AND THE HAHN-BANACH EXTENSION PROPERTY

**1.1. DEFINITION.** Let $X$ be a locally quasiconvex space over $K$ and let $M$ be a linear subspace of $X$.

a) Let us say that $M$ has the *weak extension property* (W.E.P. in short) if every $f \in M'$ has an extension $\overline{f} \in X'$.

b) As it is usual, we say that $M$ is *weakly closed* if it is a closed subspace of $X$ endowed with the weak topology $\sigma(X, X')$ (or equivalently, if for every $x \in X - M$ there exists $f \in X'$ such that $f(M) = \{0\}$ and $f(x) \neq 0$, [25] corollary 4.8).

One might expect that there is some connection between a closed subspace having the W.E.P. or being weakly closed. Unfortunately there is in general no such connection as it is showed in [16], proposition 7 and [21], p.64.

However, in the context of polar non-archimedean locally convex spaces (see section 3 for this concept) all the examples of weakly closed subspaces given in [4] have the W.E.P. and conversely. This fact allows us to formulate the following questions:

**1.2. PROBLEMS** ([4]): Let $X$ be a polar non-archimedean locally convex space over $K$.

I. Is every closed subspace of $X$ with the W.E.P. weakly closed?

II. Does every weakly closed subspace of $X$ have the W.E.P.?

In [29] *W. Schikhof* gives a negative answer to these problems.

Despite these facts, we now show that there is a global equivalence.

**1.3. DEFINITION.** We say that a locally quasiconvex space $X$ over $K$ has the *Hahn-Banach extension property* (H.B.E.P. in short) if every closed linear subspace of $X$ has the W.E.P. (equivalently, if every linear subspace of $X$ has the W.E.P.).

Then,

**1.4. THEOREM.** *For a locally quasiconvex space $X$ over $K$ the following properties are equivalent:*

i) *$X$ has the H.B.E.P.*

ii) *Every closed subspace of $X$ is weakly closed.*

iii) *For every continuous seminorm $p$ on $X$, $\ker p$ is a weakly closed subspace of $X$.*

*Proof.* The equivalence i) $\iff$ ii) was proved in theorem 4.2 of [4] in the context of non-archimedean locally convex spaces. The same proof also works here.
ii) $\Rightarrow$ iii): It is obvious.
iii) $\Rightarrow$ ii): Let $M$ be a closed subspace of $X$ and let $x_0 \in X - M$. Since $M$ is closed, there exists a continuous seminorm $p$ on $X$ and $\varepsilon > 0$ such that $M \cap \{x \in X : p(x - x_0) < \varepsilon\} = \emptyset$. Then, it is very easy to see that

$$C = \bigcup_{y \in M} \{x \in X : p(x - y) < \varepsilon\}$$

is an open absorbent and quasiconvex subset of $X$ such that $M \subset C$ and $x_0 \notin C$. Thus, the calibre $q_C$ is a continuous seminorm on $X$ for which $q_C(x_0) \geq 1$ and $q_C(M) = \{0\}$. Hence by iii) we are done.

## 2. THE HAHN-BANACH EXTENSION PROPERTY FOR $A$-NORMED SPACES

In [30] *J. van Gisbergen* proved that the spaces $\ell^q$ ($q \geq 1$) possess proper closed weakly dense subspaces. By theorem 1.4 this in turn implies that they do not have the H.B.E.P.

Neither the spaces $L^q([0,1], K)$ have the H.B.E.P., since they have trivial dual (see [32], p.89).

Also, *F. Fuentes* and *F.L. Hernández* proved ([6], theorem 2) that under certain conditions the Orlicz spaces $L^\Phi(\Omega, \mu, E)$ have trivial dual.

On the other hand *J. Martínez-Maurica* and the author showed in [12] that if $X$ is an $A$-normed space for which there exists a $t \in (0,1)$ such that $X$ has the $t$-bounded extension property (i.e., for every linear subspace $M$ of $X$ and for every $f \in M'$ there exists a continuous linear extension $\overline{f} \in X'$ verifying $t\|\overline{f}\| \leq \|f\|$), then the topology on $X$ is defined by a non-archimedean norm.

Motivated by these facts it seems natural to formulate the following problem:

**2.1. PROBLEM:** If $X$ is an $A$-normed space with the H.B.E.P., is the topology on $X$ defined by a non-archimedean norm? (or equivalently, is $X$ a non-archimedean locally convex space? [10]).

The next result gives some partial answers to this problem.

**2.2. THEOREM.** *Let $X$ be an A-normed space over $K$ with the H.B.E.P. Assume that:*

i) ([21], theorem 1.6.12) *$X$ is separable and $K$ is locally compact*

*or*

ii) ([12], theorem 5) *$X$ has a countable (Schauder) basis,*

*then the topology on $X$ is defined by a non-archimedean norm.*

### 2.3. Remarks

**1.** Recently I proved that *it is sufficient to solve problem 2.1 for A-normed spaces of countable type* (the concept of space of countable type here is as in [32]).

Indeed, suppose that the problem has an affirmative answer for spaces of countable type and let $X$ be an arbitrary $A$-normed space with the H.B.E.P. Let $(a_n)_{n\in N}$ be a weakly convergent sequence in $X$. Let $D$ be the closed linear hull of $\{a_n : n \in N\}$. Since $D$ has the W.E.P. we have by theorem 2.3 of [4] that $(a_n)_{n\in N}$ is $\sigma(D, D')$-convergent. But $D$ is a space of countable type with the H.B.E.P., which implies that $D$ is a non-archimedean locally convex space. Hence $(a_n)_{n\in N}$ is a convergent sequence ([25], proposition 4.11). By theorem 3 of [12], the conclusion follows.

By taking into account this Remark in conjunction to theorem 2.2.i) we conclude,

**PROPOSITION.** *Let $X$ be an A-normed space over a locally compact field $K$. If $X$ has the H.B.E.P., then the topology on $X$ is defined by a non-archimedean norm.*

**2.** One could ask if every $A$-normed space of countable type with separating dual has a basis. Unfortunately there is not an affirmative answer to this question, as we show in [16], corollary 8. So we derive a strong difference with respect to the well known fact (see [32], theorem 3.16) that every non-archimedean normed space of countable type has for every $t \in (0, 1)$ a $t$-orthogonal basis.

## 3. THE HAHN-BANACH EXTENSION PROPERTY FOR NON-ARCHIMEDEAN LOCALLY CONVEX SPACES

*In the sequel $X$ will denote a Hausdorff non-archimedean locally convex space over $K$.*

*From now on the term "continuous seminorm" will mean "non-archimedean continuous seminorm".*

*W. Schikhof* introduced in [25] certain classes of spaces over non spherically (i.e. non-maximally) complete fields $K$ for which a satisfactory duality theory exists. Mainly, he considered the following classes of spaces:

*Polar spaces.* $X$ is polar if its topology is defined by a family of polar seminorms. The assumption that $X$ is polar assures us of the existence of a topological dual $X'$ which separates the points of $X$ ([25], proposition 5.6).

*Strongly polar spaces.* $X$ is strongly polar if every continuous seminorm on $X$ is polar. *W. Schikhof* proved that strong polarness of $X$ is equivalent to each of the following properties:

i) (Hahn-Banach property). For each linear subspace $D$ of $X$, for each continuous seminorm $p$, for each $\varepsilon > 0$, for each $f \in D'$ with $|f| \leq p$ on $D$ there is an extension $\overline{f} \in X'$ of $f$ such that $|\overline{f}| \leq (1+\varepsilon)p$ on $X$.
ii) Every closed edged subset of $X$ is $\sigma(X, X')$-closed.
iii) For every continuous seminorm $p$ on $X$, $\{x \in X : p(x) \leq 1\}$ is $\sigma(X, X')$-closed.

Observe that these characterizations of strongly polar spaces are in the same vein as theorem 1.4 in this paper.

*Spaces with the Orlicz-Pettis property.* $X$ has the Orlicz-Pettis property ((O.P) in short) if every weakly convergent sequence in $X$ is convergent.

*Spaces of countable type.* $X$ is of countable type if for every continuous seminorm $p$ on $X$ the associated normed space $X/\mathrm{Ker}p$ is of countable type.

By using this notion of space of countable type we can extend the property (*) introduced by *A.C.M. van Rooij* [31] in the context of non-archimedean normed spaces, obtaining the following class of locally convex spaces:

*Spaces with the property* (*). $X$ has the property (*) if every closed subspace of countable type of $X$ has the W.E.P.

The relationship between these kinds of spaces and the spaces with the H.B.E.P. considered in this paper is summarized in the following result.

**3.1. THEOREM.**

a) *X of countable type ⇒ X strongly polar* ([25], theorem 4.4).

b) *X strongly polar ⇒ X polar with the H.B.E.P.* ([25], theorem 4.2).

c) *X has the H.B.E.P. ⇒ X has the property* $(*)$ (Obvious).

d) *X has the property* $(*)$ *⇒ X has (O.P.)* ([31], theorem 5.2).

e) *X metrizable and has (O.P.) ⇒ X polar* (if $P = \{p_n : n \in N\}$ is a family of seminorms determining the original topology $\tau$ on $X$, then $\overline{P} = \{\overline{p}_n : n \in N\}$ given by $\overline{p}_n = \sup\{|f| : f \in X^*, |f| \leq p_n\}$ is a family of seminorms on $X$ which defines a metrizable polar topology $\overline{\tau}$ on $X$ such that $\sigma(X, X') \leq \overline{\tau} \leq \tau$. Since $X$ has (O.P.) we derive that $\tau$ and $\overline{\tau}$ have the same convergent sequences. Hence $\tau = \overline{\tau}$ and therefore $\tau$ is polar).

*If K is not spherically complete, then*

f) *X polar with the H.B.E.P.* $\not\Rightarrow$ *X strongly polar* ([4], remark 1, p.259).

g) *X has the property* $(*)$ $\not\Rightarrow$ *X has the H.B.E.P.* ([29]).

h) *X metrizable and polar* $\not\Rightarrow$ *X has (O.P.) (e.g.* $\ell^\infty$*).*

**3.2. Remarks.** Suppose that $K$ is not spherically complete.

**1.** One of the open problems in non-archimedean analysis is about the converse of implication a) in theorem 3.1. It was posed by *W. Schikhof* in [25].

PROBLEM ([25], p.197). Is every strongly polar space of countable type?

**2.** On the other hand, all the examples of spaces with (O.P.) we know (see [4], [25], [31]) have the property $(*)$. This fact leads us to the following question:

PROBLEM. If $X$ has (O.P.), does it imply that $X$ has the property $(*)$?

## 4. THE ASSOCIATED NORMED SPACES

If $X$ is a non-archimedean locally convex space and $p$ is a continuous seminorm on $X$ we denote by $X_p$ the associated normed space $X/\ker p$ endowed with the natural norm.

A classical problem is to deduce properties of $X$ from analogous properties of the $X_p$. In our context we are interested in the properties:

1) *$X_p$ is of countable type for every continuous seminorm $p$ on $X$.*

2) *$X_p$ is strongly polar for every continuous seminorm $p$ on $X$.*

3) *$X_p$ has the H.B.E.P. for every continuous seminorm $p$ on $X$.*

4) *$X_p$ has the property $(*)$ for every continuous seminorm $p$ on $X$.*

5) *$X_p$ has (O.P.) for every continuous seminorm $p$ on $X$.*

6) *$X_p$ is a polar space for every continuous seminorm $p$ on $X$.*

Clearly, $X$ has property 1) if and only if $X$ is of countable type. Also, it is very easy to see that property 2) is equivalent to $X$ being strongly polar.

On the other hand, by theorem 3.1 it follows that

$$3) \Rightarrow 4) \Rightarrow 5) \Rightarrow 6).$$

Now it turns out that properties 3)-6) are equivalent. In fact, we have,

**4.1. THEOREM** ([4], theorem 5.2). *The following properties are equivalent:*

i) *$X_p$ has the H.B.E.P. for every continuous seminorm $p$ on $X$.*

ii) *$X_p$ is a polar space for every continuous seminorm $p$ on $X$.*

iii) *For every continuous seminorm $p$ on $X$ there exists a polar continuous seminorm $q$ that is equivalent to $p$.*

iv) *For each linear subspace $D$, for each continuous seminorm $p$, for each $f \in D'$ with $|f| \leq p$ on $D$ there exists a $C \geq 1$ and an extension $\overline{f} \in X'$ of $f$ with $|\overline{f}| \leq C.p$ on $X$.*

v) *For each continuous seminorm $p$ on $X$ there exists a $G \geq 1$ such that every closed subspace $M$ of countable type of $X_p$ has the $G$-extension property (i.e. $\forall f \in M'$, $\exists \overline{f} \in X_p'$ extending $f$ with $\|\overline{f}\| \leq G\|f\|$).*

*If $X$ satisfies one of these conditions, then $X$ is a polar space with the H.B.E.P.*

**4.2. Remark.** By looking at theorem 4.1, the following question

PROBLEM: Let $X$ satisfy properties i)–v) of theorem 4.1. Does it follow that $X$ is strongly polar?

## 5. ON COMPACTOIDITY IN NON-ARCHIMEDEAN LOCALLY CONVEX SPACES WITH THE HAHN-BANACH EXTENSION PROPERTY

In $p$-adic analysis the notion of precompactness loses some of its importance since, if the ground field is not locally compact, convex precompacts sets are reduced to a single point. To overcome this difficulty different convexifications of the concept of precompact and compact set have been introduced in the literature. Mainly,

**5.1. DEFINITION.** Let $A \subset X$ be an absolutely convex subset of $X$.

a) ([8]) $A$ is said to be *compactoid* if for each neighbourhood $U$ of 0 in $X$ there exists a finite set $H \subset X$ such that $A \subset U + c_0(H)$.

b) ([27]) $A$ is said to be *$c'$-compact* (also called *pure compactoid*) if in the above we may choose $H \subset A$.

c) ([28]) $A$ is said to be KM-*compactoid* if it is complete and there is a compact set $Y \subset X$ such that $A = \overline{c_0(Y)}$.

For non-archimedean locally convex spaces $X$ with the H.B.E.P. the author obtains in [23] several descriptions of compactoid sets of $X$ in terms of non-archimedean seminorms on the topological dual $X'$. More concretely, if for every bounded absolutely convex set $A \subset X$ we denote by $X'_A$ the dual space $X'$ of $X$ endowed with the corresponding non-archimedean seminorm with respect to the dual pair $\langle X, X' \rangle$ (i.e. $\|f\| = \sup_{x \in A} |f(x)|$ $(f \in X')$), then one verifies:

**5.2. THEOREM.** *Let $X$ be a non-archimedean locally convex space with the H.B.E.P. and let $A \subset X$ be a compactoid set in $X$.*

a) ([23], proposition 1.2) *$A$ is metrizable if and only if $X'_A$ is of countable type.*

b) *If in addition $A$ is complete and edged and we consider the following properties*

i) *$A$ is $c'$-compact.*

ii) *$\|f\|_A \in |K|$ $\forall f \in X'$ and each one-dimensional subspace of $X'$ has a $\|.\|_A$-ortho-complement.*

iii) *$A$ is a KM-compactoid.*

iv) *$X'_A$ is a normed space with an orthonormal basis.*

v) *$A$ is metrizable and $c'$-compact.*

vi) *$A$ is metrizable and KM-compactoid.*

vii) *$X'_A$ is a normed space with a countable orthonormal basis.*

*Then, we have that* ([23], proposition 2.3)
vii) $\Longleftrightarrow$ vi) $\Longleftrightarrow$ v) $\Rightarrow$ iv) $\Longleftrightarrow$ iii) $\Rightarrow$ ii) $\Longleftrightarrow$ i).
*If in addition $X$ has a countable basis (Schauder basis) such that $(X, \sigma(X, X'))$ and $(X', \sigma(X', X))$ are sequentially complete, then properties* i)–vii) *are equivalent.*

In 1955 *Grothendieck* [7] studied the approximation property for locally convex spaces over the real or complex field. He says that a locally convex space $Y$ over $R$ or $C$ has the ***approximation property*** if the identity map on $Y$ can be approximated by operators of finite rank uniformly on every compact subset of $Y$ (by an operator we understand a continuous linear mapping). Now, we reformulate this property by replacing the notion of compact set by the notion of compactoid set.

**5.3. DEFINITION.** $X$ is said to have the ***approximation property*** if the identity map on $X$ can be approximated by operators of finite rank uniformly on each compactoid subset of $X$.

In 1972 *De Grande-De Kimpe* [3] proved that every non-archimedean locally convex space over a spherically complete field $K$ has the approximation property. The following is an extension of this result.

**5.4. THEOREM.** *Every non-archimedean polar locally convex space $X$ has the approximation property.*

*Proof.* First suppose that $X$ is a normed space. Without loss of generality we can assume that the norm $\|.\|$ on $X$ is polar. Let $A \subset X$ be a compactoid set in $X$ and let $\varepsilon > 0$. By [32], theorem 4.37 there exists a sequence $(e_n)$ in $X$ with $\lim_{n\to\infty} e_n = 0$ such that $A \subset \overline{co(e_n)}$. Choose $n_0 \in N$ satisfying $\|e_n\| \leq t\varepsilon$ for all $n \geq n_0$. By [24], theorem 2.1, there exists a continuous linear projection $P : X \to [e_1, \ldots, e_{n_0}]$ with $\|P\| \leq t^{-1}$. Then $P$ is an operator of finite rank on $X$ for which we have that $\|x - Px\| \leq \varepsilon$ for all $x \in A$. Hence, we conclude that $X$ has the approximation property.

Now, suppose that $X$ is an arbitrary polar locally convex space. Let $P$ be a family of seminorms determining the topology on $X$. For every $p \in P$ the associated normed space $X_p$ is polar and hence it has the approximation property. With the same proof as in proposition 2 of [3], we derive that $X$ has also the approximation property.

**4.5. Remark.** Theorem 5.4 is in sharp contrast with the corresponding results in case of real or complex field. In fact, in 1973 *Enflo* [5] gave a negative answer to the problem whether or not every locally convex space over $R$ or $C$ has the approximation property.

## REFERENCES


[1] Bastero, J.: *Convexidad y acotación en los espacios vectoriales topológicos sobre cuerpos cuasivaluados.* Tesis Universidad de Zaragoza (1975).

[2] De Grande-De Kimpe, N.: *On spaces of operators between locally $K$-convex spaces.* Nederl. Akad. Wetensch. Indag. Math. Proc. Ser. A **75** = Indag. Math. **34** (1972), 113–129

[3] De Grande-De Kimpe, N.: *On the Grothendieck approximation property in non-archimedean analysis.* Nieuw Arch. Wisk. (3) **20** (1972), 242–245.

[4] De Grande-De Kimpe, N., Pérez-García, C.: *Weakly closed subspaces and the Hahn-Banach extension property in p-adic analysis.* Nederl. Akad. Wetensch. Indag. Math. **50** (1988), 253–261.

[5] Enflo, P.: *A counterexample to the approximation property in Banach spaces.* Acta Math. **130** (1973), 309–317.

[6] Fuentes, F., Hernández, F.L.: *El dual de espacios de Orlicz sobre cuerpos no arquimedianos.* Rev. Real Acad. Cienc. Exact. Fís. Natur. Madrid **75** (1981), 519–524.

[7] Grothendieck, A.: *Produits tensoriels topologiques et espaces nucleaires.* Mem. Amer. Math. Soc. **16** (1955).

[8] Gruson, L., Van der Put, M.: *Banach spaces.* Bull. Soc. Math. France, Mém. No. **39–40** (1974), 55–100.

[9] Ingleton, A.W.: *The Hahn-Banach theorem for non-archimedean valued fields.* Proc. Cambridge Philos. Soc. **48** (1952), 41–45.

[10] Martínez-Maurica, J., Pellón, T., Pérez-García, C.: *Espacios de Banach arquimedianos sobre cuerpos locales.* Publ. Sem. Mat. García de Galdeano Ser. II **95** (1986), 83–89.

[11] Martínez-Maurica, J., Pérez-García, C.: *Sucesiones básicas en espacios normados sobre cuerpos valuados.* Actas do VII Congresso do Grupo de Matemáticos de

Expressão latina, Vol II, 159–162 (1986).

[12] Martínez-Maurica, J., Pérez-García, C.: *The Hahn-Banach extension property in a class of normed spaces.* Quaestiones Math. **8** (1986), 335–341.

[13] Martínez-Maurica, J., Pérez-García, C.: *The three-space problem for a class of normed spaces.* Bull. Soc. Math. Belg. Sér. B **39** (1987), 209–214.

[14] Martínez-Maurica, J., Pérez-García, C.: *Local convexity is a three-space property for non-archimedean Fréchet spaces.* Extracta Mathematicae **2** (1) (1987), 1–3.

[15] Martínez-Maurica, J., Pérez-García, C.: *F-espacios sobre cuerpos locales: la propiedad de extensión.* Rev. Real Acad. Ciencias Exact. Fís. Natur. Madrid, **81** (1987), 669–676.

[16] Martínez-Maurica, J., Pérez-García, C.: *Non-archimedean K-spaces.* Houston J. Math. **14** (1988), 529–534.

[17] Monna, A.F.: *Over een lineaire p-adische ruimte.* Nederl. Akad. Wetensch. Verslagen, Afd. Natuurkunde **52** (1943), 74–82.

[18] Monna, A.F.: *Over zwakke en sterke convergentie in een p-adische Banach-ruimte.* Nederl. Akad. Wetensch. Verslagen, Afd. Natuurkunde **52** (1943), 207–211.

[19] Monna, A.F.: *Analyse non-archimédienne.* Springer-Verlag, Berlin-Heidelberg, 1970.

[20] Monna, A.F.: *Rapport sur la thèorie des espaces linéaires topologiques sur un corps valué non-archimédien.* Bull. Soc. Math. France, Mém. No. **39–40** (1974), 255–278.

[21] Pellón, T.: *Algunos aspectos de la teoria de espacios de Banach ultramétricos.* Tesis Universidad de Cantabria (1986).

[22] Pérez-García, C.: *El problema de la base en espacios A-normados.* In *Proceedings of the Eleventh Spanish-Portuguese Conference on Mathematics*, Vol. 1 (Badajoz, 1986), 375–380, Publ. Dept. Mat. Univ. Extremadura, 18, Univ. Extremadura, Badajos, 1987.

[23] Pérez-García, C.: *Compactoid sets in p-adic locally convex spaces with the Hahn-Banach extension property.* Ann. Soc. Sci. Bruxelles Sér. I **102** (1988), 97–104.

[24] Pérez-García, C.: *Semi-Fredholm operators and the Calkin algebra in p-adic analysis I-II.* Bull. Soc. Math. Belg. Sér. B **42** (1990), 69–101.

[25] Schikhof, W.: *Locally convex spaces over non-spherically complete valued fields*

*I-II*. Bull. Soc. Math. Belg. Sér. B **38** (1986), 187–224.

[26] Schikhof, W.: *Topological stability of p-adic compactoids under continuous injections.* Report 8644, Dept. Math., Catholic University, Nijmegen, The Netherlands, (1986).

[27] Schikhof, W.: *A complementary variant of c-compactness in p-adic functional analysis.* Report 8647, Dept. Math., Catholic University, Nijmegen, The Netherlands, (1986).

[28] Schikhof, W.: *Compact-like sets in non-archimedean functional analysis.* In *Proceedings of the confererence on p-adic analysis (Houthalen, 1986)*, edited by N. De Grande-De Kimpe and L. van Hamme, Vrije Universiteit Brussel, Brussels (1987), 137–147.

[29] Schikhof, W.: *The complementation property of $\ell^\infty$ in p-adic Banach spaces.* In *p-adic Analysis. Proceedings, Trento 1989*, edited by F. Baldassarri, S. Bosch and B. Dwork, Springer-Verlag (Lect. Notes 1454), Berlin-Heidelberg-New York (1990), 342–350.

[30] Van Gisbergen, J.: *Archimedisch genormeerde vectorruimten over een niet-Archimedisch gewaardeerd lichaam.* Master Thesis, Dept. Math., Catholic University, Nijmegen, The Netherlands (1985).

[31] Van Rooij, A.C.M.: *Notes on p-adic Banach spaces.* Report 7633, Dept. Math., Catholic University, Nijmegen, The Netherlands, (1976).

[32] Van Rooij, A.C.M.: *Non-archimedean Functional Analysis.* Marcel Dekker, New York, 1978.

[33] Van Tiel, J.: *Espaces localement K-convexes I-III.* Nederl. Akad. Wetensch. Proc. Ser. A **68** = Indag. Math. **27** (1965), 249–289.

Departamento de Matemáticas
Facultad de Ciencias, Universidad de Cantabria
Av. de los Castros
39071 Santander, Spain

# BANACH ALGEBRA OF $p$-ADIC VALUED ALMOST PERIODIC FUNCTIONS

G. Rangan and M.S. Saleemullah

## 1. INTRODUCTION

$p$-adic valued almost periodic functions were introduced by Rangan [1], [2] and [3] and their properties such as the existence of mean for such functions were also studied. Analogous to the classical case it was shown that the continuous almost periodic functions on a totally disconnected topological group $G$ (not necessarily locally-compact and throughout this paper $G$ stands for such a group) are the same as the continuous functions on the non-archimedean Bohr compactification $\beta(G)$ of $G$. $A = A_C(G)$ of all continuous almost periodic functions on $G$ form a Banach algebra with convolution $(*)$ as multiplication. In this paper we show that the characteristic functions $\chi_H$ of closed normal subgroups $H$ of finite index when suitably normalised give rise to the functions $U_H$ which are central idempotents in the algebra $A$. $A * U_H$ is then a finite-dimensional two sided ideal of $A$. The direct sum $\sum U_H * A$, as $H$ varies over the closed normal subgroups of finite indices is shown to be dense in the algebra $A$. When the mean exists for continuous almost periodic functions on a group $G$ then any continuous almost periodic functions on $G$ can be uniformly approximated by linear combinations of matrix coefficients of the finite-dimensional representations on $G$. We have made at the end of the paper a conjecture regarding the structure of the algebra $A$ which can possibly be proved affirmatively using the results on idempotents of $A$ in section 4 of this paper or using the structure theory of the group algebra of the compact group $\beta(G)$ of $G$.

Schikhof [5] showed that the definition of almost periodicity as given in [1], is too restrictive in the sense that even characters may not be almost periodic. To include such functions also in the class of almost periodic functions, he generalized the notion of almost periodicity using compactoids and studied the existence of mean and other properties of such functions. He showed in particular that when the mean exists on $G$ and the base field is algebraically closed, then the characters form an orthonormal base for the space $AP(G \to K)$ of almost periodic functions in the sense of Schikhof.

A number of questions such as (i) when one can extend a mean on $A_C(G)$ to $AP(G \to K)$ (ii) the structure of the algebra $AP(G \to K)$ and the nature of the subalgebra $A_C(G)$, etc. arise. We hope to take up these studies in our next paper.

## 2. NOTATIONS AND PRELIMINARIES

Let $K$ denote a complete non-archimedean non-trivial valued field of characteristic zero. We recall the following from Rangan [1] for convenience. A function $f$ defined on a group $G$ is called almost periodic if for every $\varepsilon > 0$ there exist a covering of $G$ by subsets $A_1, A_2, \ldots, A_n$ such that for $x, y \in A_i$, $i = 1, 2, \ldots, n$ and for all $c, d \in G$, $|f(cxd) - f(cyd)| < \varepsilon$. It is not difficult to see that this condition is equivalent to the fact that the left (or right) translates of $f$ form a precompact subset in the $K$-Banach space $B(G \to K)$ of all bounded functions $G \to K$ with the supremum norm $\| \quad \|_\infty$. Interestingly it turns out that for a given $\varepsilon > 0$ and an almost periodic function $f$ on $G$ the covering consisting of minimum number of subsets $A_1, A_2, \ldots, A_n$ such that for $x, y \in A_i$, $|f(cxd) - f(cyd)| < \varepsilon$, $i = 1, 2, \ldots, n$ is the covering by cosets of a suitable normal subgroup of finite index in $G$. This normal subgroup $H(f, \varepsilon) = H$ is called the $\varepsilon$-kernel of $f$. When $G$ is a topological group and $f$ is a continuous almost periodic function on $G$, this $\varepsilon$-kernel is easily seen to be an open and closed normal subgroup of finite index. A topological group $G$ is called and Index Bounded group or in short IB-group if $\inf |n| > 0$, $n$ varying over the indices of normal subgroups of finite index and for such a group we take $k = \inf |n|$. The mean exists for every continuous almost periodic function on $G$ if and only if $G$ is an IB-group (Theorem 3.2, Rangan [3]). The space $A_C(G)$ of all continuous almost periodic functions on an IB-group $G$ is a Banach algebra with convolution as multiplication and with respect to the norm defined by $\|f\| = \frac{1}{k} \sup |f(x)|$, $x \in G$. In what follows we use the notations of [1] and [2] and $G$ always denote a totally disconnected topological group unless otherwise specified.

## 3. PETER-WEYL-VON NEUMANN THEOREM

THEOREM 3.1. *Let $G$ be a totally disconnected topological group on which the mean exists for continuous almost periodic functions. Then each closed normal subgroup of finite index gives rise to a finite-dimensional representation of the group $G$. The matrix coefficient of these representations are the characteristic functions of the cosets of this subgroup. Any continuous almost periodic function can be uniformly approximated by linear combinations of the matrix coefficients of the finite-dimensional representations defined by its $\varepsilon$-kernels.*

*Proof.* Let $H$ be a closed normal subgroup of finite index $n$ in $G$. Let $h_1, h_2, \ldots, h_n$ be the characteristic functions of the left cosets of $H$ in $G$. Let $L$ be the finite-dimensional subspace generated by them in $A_C(G)$. Then $L$ is an invariant subspace of dimension $n$. The right translation operator on $L$ gives rise to a representation $D : s \to D(S)$ where $D(s)$ is an $n \times n$-matrix, $D(s)$ acting on $L$ as a linear transformation. If $a_{ij}(s)$ are the coefficients of this representation then

$$(h_i)_s(t) = h_i(ts) = \sum_{j=1}^{n} a_{ij}(s)h_j(t).$$

Taking $t = e$, the identity of $G$, we see that

$$h_i(s) = \sum_{j=1}^{n} a_{ij}(s)h_j(e).$$

If $h_1$ is the characteristic function of $H$, then $h_i(s) = a_{i1}(s)$. Thus we see that $a_{i1}$ is the characteristic function $h_i$ for $i = 1, 2, \ldots, n$. Similarly we can show that each $a_{ij}$ is one of $h_1, h_2, \ldots, h_n$. Clearly $a_{ij}$'s are bounded functions and so $D$ is a bounded representation. Theorem 7.4 [1] states that any continuous almost periodic function $f$ can be uniformly approximated by finite linear combinations of the characteristic functions of the left (right) cosets of the $\varepsilon$-kernels of $f$. Hence taking $H = H(f, \varepsilon)$ in the above we see that the $a_{ij}$'s are these characteristic functions.

## 4. ALGEBRA $A_C(G)$ OF CONTINUOUS ALMOST PERIODIC FUNCTIONS

PROPOSITION 4.1. *Let $H$ be a closed normal subgroup of finite index $n$ in $G$ and $U_H = \frac{1}{M(\chi_H)}\chi_H = n\chi_H$. Then $U_H$ belongs to the centre of $A_C(G)$.*

*Proof.* Since $H$ is normal,

$$\begin{aligned} U_H * f(x) &= \frac{1}{M(\chi_H)} M_y(\chi_H(xy^{-1})f(y)) \\ &= \frac{1}{M(\chi_H)} M_y(\chi_H(yx^{-1})f(y)) \\ &= M_y(U_H(yx^{-1})f(y)) \\ &= f * U_H(x). \end{aligned}$$

**PROPOSITION 4.2.** *Put* $f_H(x) = M_y^H(f(xy))$ *where* $f \in A_C(G)$. *Then* $f * U_H = U_H * f = f_H$.

*Proof.* Using left invariance of the mean we have

$$\begin{aligned} f * U_H(x) &= M_y(f(y)U_H(y^{-1}x)) \\ &= M_y(f(y)U_H(x^{-1}y)) \\ &= M_y(f(xy)U_H(y)) \\ &= \frac{1}{M(\chi_H)} M_y(f(xy)\chi_H(y)) \\ &= M_y^H(f(xy)) \end{aligned}$$

since $M^H(f_x \mid H) = \frac{1}{M(\chi_H)} M({}_xf \cdot \chi_H)$ where ${}_xf(y) = f(xy)$.

**PROPOSITION 4.3.** $f_H$ *is constant on the cosets of* $H$.

*Proof.* Let $xx'^{-1} = h \in H$. Then

$$f_H(x) = M_y^H(f(xy)) = M_y^H(f(xx^{-1}x'y) = M_y^H(f(x'y)) = f_H(x'),$$

where the second inequality follows by translation invariance of $M_y^H$, since $x^{-1}x' = h \in H$.

**COROLLARY 4.4.** *If* $f \in A_C(G)$ *and* $f$ *is constant on the left cosets of* $H$ *then*

$$f * U_H = U_H * f = f = f_H.$$

*Proof.*

$$f_H(x) = f * U_H(x) = M_y^H(f(xy)) = M_y^H(f(x)) = f(x),$$

since $f(xy) = f(x)$ for $y \in H$.

**COROLLARY 4.5.** *$U_H$ is a central idempotent.*

*Proof.* By Proposition 4.1, it is enough to prove that $U_H$ is idempotent which follows from corollary 4.4, since $U_H$ is constant on the left cosets of $H$.

**PROPOSITION 4.6.** $M_{G/H}(f_H) = M(f)$.

*Proof.* $f_H$ is constant on the cosets of $H$ in $G$ and hence can be considered as a function on the finite discrete group $G/H$. Now, since $M(\chi_H) = 1/n$ we have

$$\begin{aligned}
M_{G/H}(f_H) &= \frac{1}{n}\sum_{xH} f_H(xH) = \frac{1}{n}\sum_{xH} f(x) \\
&= \frac{1}{n}\sum_{xH} M_y(f(y)U_H(x^{-1}y)) \\
&= M(\chi_H)\sum_{xH} M_y(f(y)U_H(x^{-1}y)) \\
&= \sum_{xH} M_y(f(y)\chi_H(x^{-1}y)) \\
&= \sum_{xH} M_y(f(y)\chi_{xH}(y)) \\
&= M_y(f(y)) = M(f).
\end{aligned}$$

**PROPOSITION 4.7.** $\|U_H\| \geq 1$.

*Proof.* $|G/H| = n$. If $\ell$ is the index of a normal subgroup of finite index in $G/H$, then $n = \ell\ell'$ for some $\ell' \geq 1$. So $|n| = |\ell\ell'| = |\ell|\ |\ell'| \leq |\ell|$ since $|\ell'| \leq 1$. But the trivial subgroup $\{H\}$ in $G/H$ has index $n$ itself. $k(G/H) = \inf |\ell| = |n|$, $\ell$ varying over the indices of normal subgroups of finite indices in $G/H$. Hence

$$\begin{aligned}
\|U_H\| &= \frac{1}{k}\sup|U_H(x)| = \frac{1}{k}\sup|n|\ |\chi_H(x)| \\
&= \frac{|n|}{k} \geq 1.
\end{aligned}$$

It is not clear whether $\|U_H\|$ is actually equal to 1.

Consider the collection $\{U_H\}$, $H \in \Gamma'_G$ where $\Gamma'_G$ denotes the collection of all closed normal subgroups of finite indices.

$A_C(G) * U_H$ is a two sided ideal of $A_C(G)$ and $f * U_H$, being a function constant on the cosets of $H$, can be identified with a function in $A_C(G/H)$. It is clear that $A_C(G/H)$ coincides with the vector space of all functions on the finite group $G/H$ and so as a vector space $A_C(G/H)$ is isomorphic with ${}_K[G : H]$. Thus $A * U_H$ can be identified with the finite-dimensional space ${}_K[G : H]$.

In the collection of all non-zero central idempotents in $A = A_C(G)$, we introduce an order as follows: If $e_1, e_2$ are non-zero central idempotents then $e_1 \leq e_2$ if $e_1 * A \subset e_2 * A$. Let $E$ be the set of all non-zero minimal central idempotents of $A$.

**PROPOSITION 4.8.** *Corresponding to every $e \in E$, there exists an $H \in \Gamma'_G$ such that $e * U_H = e$.*

*Proof.* Note that for every $e \in E$, $\|e\| \geq 1$.
For $\|e\| = \|e * e\| \leq \|e\| \, \|e\|$, i.e. $\|e\|(\|e\| - 1) \geq 0$.

$\|e\| > 0$ now implies $\|e\| \geq 1$. (This also gives an alternative way of seeing that $\|U_H\| \geq 1$.)
Let $0 < k = k(G) =$ the index of $G$. Choose $\varepsilon > 0$ and $\varepsilon < k$. Corresponding to this $\varepsilon > 0$, there exists $H \in \Gamma_G$ (see the proof of Lemma 6.2, [1]) such that

$$|f(x) - f * U_H(x)| < \varepsilon \text{ for all } x \in G.$$

Hence $\|f - f * U_H\| = \frac{1}{k} \sup |f(x) - f * U_H(x)| \leq \varepsilon/k < 1$.
Taking $f = e$ we get $\|e - e * U_H\| < 1$.
If $e * U_H = 0$, then $\|e\| < 1$, a contradiction. Therefore $e * U_H \neq 0$, and it is a central idempotent since $e$ and $U_H$ are so. Further $e * U_H * A \subset e * A$ since $U_H * A \subset A$ (i.e.) $e * U_H \leq e$. The minimality of $e$ now implies that $e * U_H = e$.

**PROPOSITION 4.9.** *There are only a finite number of non-zero minimal central idempotents $\leq U_H$ for every $H \in \Gamma'_G$. If for any fixed $H$, $e_1, e_2 \ldots e_n$ are such idempotents then $U_H = e_1 + e_2 + \cdots + e_n$.*

*Proof.* $(e_1 + e_2 + \cdots + e_n) * (e_1 + e_2 \cdots + e_n) = e_1 + e_2 \cdots + e_n$ since $e_i * e_j = 0$ for $i \neq j$ and $e_i * e_i = e_i$ i.e. $(e_1 + e_2 \cdots + e_n)$ is an idempotent and it is clearly central since each $e_i$ is so. Further

$$\begin{aligned}(e_1 + e_2 + \cdots + e_n) * A &= e_1 * A + \cdots + e_n * A. \\ &\subset U_H * A + U_H * A + \cdots + U_H * A \\ &\subset U_H * A \text{ (since } U_H * A \text{ is a subspace)}\end{aligned}$$

(i.e.) $(e_1 + e_2 + \cdots + e_n) \leq U_H$:

$$e_1 + e_2 + \cdots + e_n \neq 0$$

since otherwise

$$e_1 = e_1 * (e_1 + \cdots + e_n) = 0,$$

a contradiction.

Consider $e = U_H - (e_1 + e_2 + \cdots + e_n)$

$$\begin{aligned} e * e &= [U_H - (e_1 + e_2 + \cdots + e_n)] * [U_H - (e_1 + \cdots + e)] \\ &= U_H * U_H - U_H * (e_1 + \cdots + e_n) \\ &\quad - (e_1 + e_2 \cdots + e_n) * U_H + (e_1 + \cdots + e_n) * (e_1 + \cdots + e_n) \end{aligned}$$

$U_H * e_i * A \subset U_H * A$ and $e_i * A$, and so $U_H * e_i \leq U_H$ and $U_H * e_i \leq e_i$.

But $e_i$'s are minimal idempotents and so $U_H * e_i = e_i$ for $i = 1, 2, \ldots, n$. Hence $e * e = U_H - (e_1 + \cdots + e_n) = e$ (i.e.) $e$ is an idempotent and clearly $e$ is central

$$\begin{aligned} e * A &= U_H * A - (e_1 + e_2 + \cdots + e_n) * A \\ &\subset U_H * A \end{aligned}$$

since $(e_1 + e_2 + \cdots + e_n) * A \subset U_H * A$, and $U_H * A$ is a subspace.

If $e \neq 0$, then $e \leq U_H$ and from the definition of the $e_i$'s it follows that $e = e_i$ for some $i$. Then $e_i = e = U_H - (e_1 + \cdots + e_n)$ (i.e.) $U_H = e_1 + \cdots + 2e_i + \cdots + e_n$

$$e_i = U_H * e_i = (e_1 + \cdots + 2e_i + \cdots + e_n) * e_i = 2e_i * e_i = 2e_i$$

(i.e.) $e_i = 0$ a contradiction.

Hence $e = 0$ (i.e.) $U_H = (e_1 + e_2 + \cdots + e_n)$.

**PROPOSITION 4.10.** *For $e \in E$, $e * A$ is a finite-dimensional two sided ideal.*

*Proof.* Corresponding to $e \in E$, $\exists H \in \Gamma'_G$ such that

$$e * U_H = e (= U_H * e).$$

Hence $e * A = e * U_H * A \subset U_H * A = U_H * e * A$ (since $e * A \subset A$).

$U_H * A$ is a f.d. space $\Longrightarrow$ $e * A$ is finite-dimensional.

Since $e$ is a central idempotent $e * A$ is a two sided ideal of $A$.

COROLLARY 4.11. *$e * A \simeq K^r$ for some $r$.*

*Proof.* $e * A$ is a finite-dimensional subspace of $U_H * A$ for some $H$, $U_H * A \simeq {}_K[G : H]$ now implies by Lemma 2.1 [5] that $e * A \simeq K^r$ for some $r \leq [G : H]$.

LEMMA 4.12. *If $f \in A$, $\varepsilon > 0$, then there exists a closed normal subgroup $H$ of finite index such that for all closed normal subgroups $H'$ of finite-index such that $H' \subset H$*

$$|f(x) - f * U_{H'}(x)| \leq \varepsilon,$$

*where $U_{H'}(x) = \chi_{H'/M(\chi_{H'})}$.*

*Proof.* Since $f$ is almost-periodic let $H = H(f, k\varepsilon)$ be the $k\varepsilon$-kernel of $f$. Then for $x \in G$, $y \in H(f, k\varepsilon) = H$, $|f(x) - f(xy^{-1})| < k\varepsilon$. Let $H' \subset H$, $H'$ a closed normal subgroup of finite index in $G$. Put $U_{H'} = \chi_{H'/M(\chi_{H'})}$, $M(U_{H'}) = 1$.

$$\begin{aligned}|f(x) - f * U_{H'}(x)| &= |f(x)M_y(U_{H'}(y)) - M_y(f(xy^{-1})U_{H'}(Y))| \\ &= |M_y\{(f(x) - f(xy^{-1}))U_{H'}(y)\}| = M_y\{\phi_x(y)\}\end{aligned}$$

where $\phi_x(y) = (f(x) - f(xy^{-1}))U_{H'}(y)$.
By Proposition 4.3, Rangan [1]

$$|M_y(\phi_x(y))| \leq \|\phi_x\|/k.$$

Let $M_y(\chi_{H'}(y)) = 1/n'$ where $[G : H'] = n'$.
For $y \in H'$ and any $x \in G$, since $H' \subset H$

$$\begin{aligned}|\phi_x(y)| &= |f(x) - f(xy^{-1})|\,|U_{H'}(y)| \\ &< k\varepsilon \cdot |\chi_{H'}(y)|/|M(\chi_{H'})| \\ &< k\varepsilon|n'| \leq k\varepsilon.\end{aligned}$$

For $y \notin H'$, $\phi_x(y) = 0$

$$\|\phi_x\| = \sup_{y \in G} |\phi_x(y)| \leq k\varepsilon \quad \text{for all } x \in G.$$

Consequently

$$|f(x) - f * U_{H'}(x)| = |M_y(\phi_x(y))|\,\frac{\|\phi_x\|}{k} \leq \varepsilon.$$

COROLLARY 4.13. *$\lim_{H \in \Gamma'_G} f * U_H = f$, i.e. $\{U_H\}_{H \in \Gamma'_G}$ form an approximate identity for $A$.*

**THEOREM 4.14.** $\sum_{H \in \Gamma'_G} U_H * A$ *dense in* $A$.

*Proof.* Clear from the previous corollary.

It is known (see Van Rooij [4]) that for a compact $p$-free group, if $E$ denotes the set of all minimal central idempotents of $L(G)$, the group algebra of $G$, $B_e = L(G) * e$ is a finite dimensional non-zero two sided ideal in $L(G)$ and every non-zero minimal two sided ideal of $L(G)$ is a $B_e$ for a suitable $e \in E$. If $I$ is a closed two sided ideal in $L(G)$ then $I = \text{clo} \sum_{e \in I} B_e$. For every $f \in L(G)$, (where clo means the closure)

$$f = \sum_{e \in E} e * f \text{ and } \|f\| = \sup_{e \in E} \|e * f\|.$$

Further there exists a Banach algebra isomorphism between

$$L(G) \quad \text{and} \quad \bigoplus_{e \in E} B_e.$$

Conjecture: It seems that one can get a similar result for the Banach algebra $A_C(G)$ when $G$ is an Index bounded group.

Possibly one can try to prove this using the above theorem 4.14 and the properties of minimal central idempotents proved in this section or using the structure theory of the group algebra of the Bohr compactification $\beta(G)$ which is a compact group.

## REFERENCES

[1] Rangan, G.: *Non-archimedean valued almost periodic functions.* Nederl. Akad. Wetensch. Proc. Ser. A **72** = Indag. Math. **31** (1969), 345–353.

[2] Rangan, G.: *Non-archimedean Bohr compactification of a topological group.* Nederl. Akad. Wetensch. Proc. Ser. A **72** = Indag. Math. **31** (1969), 354–360.

[3] Rangan, G.: *On the existence of non-archimedean valued invariant means.* Publ. Math. Debrecen **29** (1982), 57–63.

[4] Van Rooij, A.C.M.: *Non-Archimedean Functional Analysis.* Marcel Dekker, New York, 1978.

[5] Van Rooij, A.C.M. and W.H. Schikhof: *Group representations in non-archimedean Banach spaces.* Bull. Soc. Math. France, Mém. No. **39–40** (1974), 329–340.

[6] Schikhof, W.H.: *An approach to p-adic almost periodicity by means of compactoids*. Report 8809, Dept. Math., Catholic University, Nijmegen, The Netherlands (1987).

THE RAMANUJAN INSTITUTE
UNIVERSITY OF MADRAS
MADRAS - 5, INDIA

# THE AXIOM OF CHOICE IN $p$-ADIC FUNCTIONAL ANALYSIS

A.C.M. van Rooij

Functional analysts generally are little concerned with the set-theoretic background of their theories. This lecture* is an attempt to evoke interest in the role of the Axiom of Choice (**AC**) in $p$-adic Functional Analysis.

It is a deplorably wide-spread attitude with mathematicians in general and with functional analysts in particular blindly to accept the Axiom of Choice without thinking of its cost. The Axiom enables one to claim the existence of certain objects but only if one manages not to think about what "existence" means. It is a magic key to open a door that is closed to constructivists, but the door leads to a phantom world of things one cannot touch.

Let me hasten to add that the Countable Axiom of Choice is a quite different matter. It can be understood, it can be made constructive and, best of all, it seems to be everything one needs, at least in Functional Analysis over $\mathbb{R}$ and $\mathbb{C}$. For instance, it implies the Hahn-Banach Theorem for separable normed spaces. Admittedly it does not yield the Hahn-Banach Theorem for $\ell^\infty$, but does anyone ever use that?

The dubious value of the full Axiom of Choice becomes apparent if one considers alternative axioms, just as valid, that lead to a theory in which all subsets of $\mathbb{R}$ are Lebesgue measurable, all linear functionals on Banach spaces are continuous and all finitely additive measures on a $\sigma$-algebra are $\sigma$-additive. Such a theory is in no sense inferior; it is just unfashionable. Next year may be different.

It is not my purpose to abolish **AC**. (After all, it may well be "true", if that means anything.) I do wish that mathematicians were more aware of it and did not take it for granted.

As I said before, the Countable Axiom of Choice seems to suffice for Functional Analysis over $\mathbb{R}$ and $\mathbb{C}$. It may be more crucial in the $p$-adic theory. Indeed, it looks

* This lecture, as given at the Conference, was titled "Open Problems in $p$-adic Functional Analysis" and consisted of two parts. The first is presented here, the second is subsumed in the paper "Open Problems" elsewhere in these Proceedings.

as if from the point of view of the set-theoretic foundations, there is a big difference between classical and $p$-adic Functional Analysis. As a case in point I would like to discuss the Hahn-Banach Theorem and its $p$-adic counterpart, Ingleton's Theorem.

The standard proofs of the Hahn-Banach Theorem (**HBT**) one encounters in literature rely on the Axiom of Choice (**AC**) and, indeed, it is well known that the theorem does not follow from the Zermelo-Fraenkl axioms for Set Theory. On the other hand it does not require the full strength of the Axiom of Choice, either. A weaker axiom implying **HBT** is the Ultrafilter Theorem (**UT**), stated farther on. Assuming the Zermelo-Fraenkl axioms one has

$$\mathbf{AC} \Rightarrow \mathbf{UT} \Rightarrow \mathbf{HBT}.$$

It can be proved that the reverse implications are false:

$$\mathbf{AC} \not\Leftarrow \mathbf{UT} \not\Leftarrow \mathbf{HBT}.$$

We propose to show how *for locally compact $K$*, Ingleton's Theorem can be deduced from **UT**. Our proof is only a trivial adaption of Luxemburg's proof of **HBT**. (See [1].)

Throughout, we base ourselves on the Zermelo-Fraenkl axioms. We do not use other set-theoretic axioms without explicit mention.

As usual, $K$ is a field provided with a nontrivial non-Archimedean valuation. In addition we assume that $K$ is locally compact

**§1.** First, the Ultrafilter Theorem itself.

Let $T$ be a nonempty set. A nonempty collection $\mathcal{C}$ of subsets of $T$ is said to have the *finite intersection property* if

$$N \in \mathbb{N},\ X_1, \ldots X_N \in \mathcal{C} \implies X_1 \cap \ldots \cap X_N \neq \varnothing.$$

(Then in particular $\varnothing \notin \mathcal{C}$.)

An *ultrafilter* in $T$ is a collection of subsets of $T$ that is maximal among the collections that have the finite intersection property. E.g., for every $\tau_0 \in T$, $C := \{X : \tau_0 \in X\}$ is an ultrafilter.

The *Ultrafilter Theorem* states:

$$\textbf{UT}\quad\left[\begin{array}{l}\text{Let } T \text{ be a nonempty set, } \mathcal{C} \text{ a}\\ \text{nonempty collection of subsets of } T\\ \text{having the finite intersection property.}\\ \text{Then } \mathcal{C} \text{ is contained in an ultrafilter.}\end{array}\right.$$

Clearly, **UT** is a special case of Zorn's Lemma, hence follows from the Axiom of Choice. The reverse is false:

$$\textbf{AC} \not\Leftarrow \textbf{UT}.$$

Various theorems in mathematics that commonly are obtained from **AC** actually follow already from **UT**. An example of this is the classical Gelfand-Naimark-Segal representation theorem for commutative $C^*$-algebras. Another is the Alaoglu Theorem: **AC** is known to be equivalent to the Tychonoff Theorem, but **UT** is enough to imply compactness of $[0,1]^S$ for every $S$, and thereby the Alaoglu Theorem (but not the Krein-Milman Theorem).

§**2.** Let $\mathcal{U}$ be an ultrafilter in a set $T$. Then:
(1) $T \in \mathcal{U}$. More generally: if $X \in \mathcal{U}$ and $X \subset Y \subset T$, then $Y \in \mathcal{U}$.
(2) If $X \in \mathcal{U}$ and $Y \in \mathcal{U}$, then $X \cap Y \in \mathcal{U}$.
(3) If $X \cup Y \in \mathcal{U}$, then $X \in \mathcal{U}$ or $Y \in \mathcal{U}$.

The proofs of these statements are elementary.

The following terminology will be convenient for our proof of Ingleton's Theorem from **UT**. Again, $\mathcal{U}$ is an ultrafilter in $T$.

Suppose for every $\tau \in T$ we are given a proposition $P(\tau)$. We say

$$P(\tau)\ \textit{for almost every}\ \tau$$

if $\{\tau : P(\tau)\} \in \mathcal{U}$. Instead of "almost every" we also use the abbreviation "a.e.".

Assume that for every $\tau$ we have propositions $P(\tau)$ and $Q(\tau)$. Then from (1)-(3) one easily obtains:

$$(1')\qquad \left.\begin{array}{l} P(\tau) \text{ for a.e. } \tau \\ P(\tau) \Rightarrow Q(\tau) \text{ for a.e. } \tau \end{array}\right] \Longrightarrow Q(\tau) \text{ for a.e. } \tau.$$

$$\text{(2')} \qquad \left.\begin{array}{l} P(\tau) \text{ for a.e. } \tau \\ Q(\tau) \text{ for a.e. } \tau \end{array}\right] \Longrightarrow P(\tau) \wedge Q(\tau) \text{ for a.e. } \tau.$$

$$\text{(3')} \qquad P(\tau) \vee Q(\tau) \text{ for a.e. } \tau \Longrightarrow Q(\tau) \text{ for a.e. } \tau.$$

**§3.** Now we turn to Ingleton's Theorem.

$K$ is locally compact, $E$ is a normed vector space (norm: $\| \ \|$), $D$ a linear subspace of $E$, $f$ a continuous linear function $D \to K$. Our problem is to extend $f$ to a linear function $\overline{f} : E \to K$ with

$$|\overline{f}(x)| \leq \|f\| \, \|x\| \qquad (x \in E).$$

$$\begin{array}{ccc} D & \subset & E \\ f \searrow & & \swarrow \overline{f} \\ & K & \end{array}$$

If there exist $a_1, \ldots, a_N$ in $E$ such that $D + Ka_1 + \cdots + Ka_N = E$, there is no problem: $K$ being spherically complete one can use the standard technique to extend $f$ seccessively to the spaces

$$D + Ka_1, \ D + Ka_1 + Ka_2, \ldots, \ D + Ka_1 + \cdots Ka_N.$$

In other words, for any finite dimensional linear subspace $F$ of $E$ there is a linear $g : D + F \to K$ extending $f$ and with $\|g\| \leq \|f\|$. The entire difficulty is to get from here to an extension defined on all of $E$, and this is, of course, where **UT** comes in.

**§4.** Let $T$ be the set of all pairs $(F, g)$ where $F$ is a finite dimensional linear subspace of $E$ and $g$ is a linear function $D + F \to K$ for which

$$|g(x)| \leq \|f\| \, \|x\| \qquad (x \in E + F).$$

For $\tau = (F, g) \in T$ we write $F_\tau := F$, $g_\tau := g$.

As we have already observed, for any finite dimensional linear subspace $F$ of $E$ there exists a "Hahn-Banach extension" of $f$ to $E+F$, i.e. there exists a $\tau \in T$ with $F_\tau = F$.

For $x \in E$, set

$$T_x := \{\tau \in T : x \in F_\tau\}.$$

By the above, the collection of sets $\{T_x : x \in E\}$ has the finite intersection property. The Ultrafilter Theorem claims the existence of an ultrafilter $\mathcal{U}$ in $T$ such that

$$T_x \in \mathcal{U} \qquad (x \in E).$$

Using the phrase "almost every" as in §2 we obtain:

(1) $\quad x \in E \implies x \in F_\tau$ for almost every $\tau$.

Therefore,

(2) $\quad x \in E \implies g_\tau(x)$ exists for a.e. $\tau$.

If $x, y \in E$ then for almost every $\tau$ we have $x \in F_\tau$, $y \in F_\tau$ and $x + y \in F_\tau$. Hence,

(3) $\quad x, y \in E \implies g_\tau(x) + g_\tau(y) = g_\tau(x+y)$ for a.e. $\tau$.

Similarly,

(4) $\quad x \in E,\ \lambda \in K \implies \lambda g_\tau(x) = g_\tau(\lambda x)$ for a.e. $\tau$.

Take $x \in E$.

Let $\varepsilon > 0$. $K$ being locally compact, the (bounded) set $\{g_\tau(x) : \tau \in T_x\}$ is covered by finitely many balls of radius $\varepsilon$. It follows from (3') of §2 that one of these balls must contain $g_\tau(x)$ for a.e. $\tau$.

Thus, for every $\varepsilon > 0$ there is a (unique) ball $B_\varepsilon$ of radiud $\varepsilon$ with $g_\tau(x) \in B_\varepsilon$ for a.e. $\tau$. Consequently, we have a unique element $\overline{f}(x)$ of $K$ such that for every $\varepsilon > 0$

$$|g_\tau(x) - \overline{f}(x)| < \varepsilon \quad \text{for a.e. } \tau.$$

In this way we have obtained a function $\overline{f}$ on $E$. It is a simple matter to show that $\overline{f}$ is linear, that

$$|\overline{f}(x)| \le \|f\|\,\|x\| \qquad (x \in E)$$

and that $\overline{f}$ coincides with $f$ on $D$.

**§5.** So far, no problem. All we have done is to prove a special case of Ingleton's Theorem under slightly weaker set-theoretical assumptions than is customary. Now the questions.

A. It is possible to derive the full Ingleton Theorem from **UT**? Owing to the great diversity of the admissible scalar fields $K$, Ingleton's Theorem seems to have a much wider scope than the Hahn-Banach Theorem and may well imply **AC**. By the same token, Ingleton's Theorem for locally compact scalar fields may well imply **UT**.

B. Is there any implication between Ingleton's Theorem and the Hahn-Banach Theorem itself?

C. There are other places in $p$-adic Functional Analysis where **AC** is used, such as the construction of the spherical completion. It is really needed there?

D. Van der Put's theorem stating that for every compact space $X$, $C(X)$ has an orthonormal base relies on the existence of a Hamel base in a vector space over the residue class field. It is known that, if *every* vector space has a Hamel base, then **AC** holds (Th. 6.33 in [2]). Does van der Put's theorem imply **AC**?

E. Is it possible to obtain the spherical completion of a normed vector space without esoteric tools such as **AC** ot **UT**?

## REFERENCES

[1] Luxemburg, W.A.J.: *Two applications of the method of construction by ultrapowers to analysis.* Bull. Amer. Math. Soc. **68** (1962) 416–419

[2] Rubin, H. and Rubin, J.E.: *Equivalents of the Axiom of Choice.* North Holland, Amsterdam-New York-Oxford, 1985.

KATHOLIEKE UNIVERSITEIT
MATHEMATISCH INSTITUUT
TOERNOOIVELD
6525 ED NIJMEGEN, THE NETHERLANDS

# THE EQUATION $y' = \omega y$ AND THE MEROMORPHIC PRODUCTS

Marie-Claude Sarmant and Alain Escassut

**ABSTRACT.** Let $K$ be a complete ultrametric algebraically closed field of characteristic zero and let $D$ be a clopen bounded infraconnected set whose holes have diameter greater than a positive constant $\rho$.

Let $H(D)$ be the Banach algebra of the Analytic Elements in $D$.

The Differential Equation $\ell(\omega)$ $y' = \omega y$ $(\omega \in H(D))$ has non zero solutions in $H(D)$ if and only if $\omega$ is of the form

$$(1) \qquad \sum_{i=1}^{q} \frac{u_i}{x-c_i} + \sum_{n=1}^{\infty} \left( \frac{1}{x-a_n} - \frac{1}{x-b_n} \right)$$

where the $c_i$ belong to different holes $T_i$ of $D$, each $u_i$ is the residue of $\omega$ in $T_i$ and the sequences $(a_n), (b_n)$ satisfy conditions like $\lim_{n\to\infty} (a_n - b_n) = 0$. The solutions then are in the form $\lambda \prod_{i=1}^{q} (x-c_i)^{u_i} \prod_{n=1}^{\infty} \left(\frac{x-a_n}{x-b_n}\right)$.

Then we characterize the $\omega$ which have integrals in $H(D)$ as the elements of the form $\frac{1}{p^h} \sum_{n=1}^{\infty} \frac{1}{x-a_n} - \frac{1}{x-b_n}$ and we define and study the quasi-integrable elements.

Let $B$ be the open unit disk, and let $\theta \in H(B)$. We assume $\ell(\theta)$ to have a non zero solution $f$ analytic function in $B$ (not supposed to belong to $H(B)$). Then $f^N \in H(B)$ for some $N \in \mathbb{N}$ if and only if $\theta$ is of the form $\frac{1}{N}\omega$ where $\omega$ is of the form (1) above. As a corollary we have an immediate demonstation to show that a function $\sqrt[N]{1+x}$ does not belong to $H(B)$. When $K = \mathbb{C}_p$, we apply these results to the search of a Frobenius Structure on $\ell(\omega)$.

## 1. INTRODUCTION AND GENERAL THEOREMS

Let $K$ be an algebraically closed complete ultrametric field of characteristic zero, whose absolute value is denoted by $|\cdot|$. Let $D$ be a subset of $K$, let $R(D)$ be the

$K$-algebra of the rational functions $h(x) \in K(x)$ with no pole in $D$ and let $H(D)$ be the completion of $R(D)$ for the topology of uniform convergence on $D$. The elements of $H(D)$ are called the Analytic Elements on $D$.

Let us recall that the topologies of $R(D)$ and $H(D)$ are determined by the norm of uniform convergence $\|\cdot\|_D$ if and only if $D$ is closed and bounded [$E_1, E_2$], so that $H(D)$ is a $K$-Banach algebra (with respect to the norm $\|\cdot\|_D$) if and only if $D$ is closed and bounded.

A set $D$ is said to be *infraconnected* if for every $a \in D$, the mapping defined in $D$ by $x \to |x-a|$ has an image whose closure in $\mathbb{R}$ is an interval [$E_2$]. We know that an algebra $H(D)$ has non trivial idempotents if and only if $D$ is not infraconnected [$E_1$]. Moreover, when $D$ is clopen, the relation $f' = 0$ implies $f =$ constant if and only if $D$ is infraconnected [$E_5$]. (We call a clopen set a set which is closed and open.)

After several articles dedicated to the equation $\ell(\omega)\ y' = \omega y$ with $\omega$ and $y \in H(D)$ [$E_6, E_7, E_8, E_9$] here we will study the form of $\omega$ and the solutions $y$ by putting them into Meromorphic Products thanks to the factorization obtained in [$S_4$]. We will widen and emphasize a notion of residue for $\omega$ generalizing results obtained by G. Christol when $\omega$ was a rational function [$C_1$] and we will first characterize the $\omega \in H(D)$ such that $\ell(\omega)$ has solutions invertible in $H(D)$.

For $a \in K$ and $r \in \mathbb{R}_+$ we denote by $d(a,r)$ (resp. $d(a,r^-)$) the disk $\{x \mid |x| \leq r\}$ (resp. $\{x \mid |x| < r\}$). Let $D$ be a closed bounded set of diameter $R$. We denote by $\widetilde{D}$ the disk $d(a,R)$ with $a \in D$. We know that $\widetilde{D}\backslash D$ admits a partition $(T_i)_{i\in I}$ where each $T_i$ is a disk $d(a_i, r_i^-)$ such that $r_i$ is the distance of $a_i$ to $D$ [$E_2$]. The $T_i$ are called *the holes* of $D$.

Theorem 1 requires a notion of residue linked to a hole for an Analytic Element. This is given in Lemma 1. When $D$ is a non bounded set we will denote by $H_0(D)$ the set of the $f \in H(D)$ such that $\lim\limits_{\substack{|x|\to\infty \\ x\in D}} f(x) = 0$.

We still have to recall the Mittag-Leffler Theorem [K,R].

*Let $D$ be a closed bounded infraconnected set and let $f \in H(D)$. There exists a unique sequence $(T_n)_{n\in\mathbb{N}^*}$ of holes of $D$ and a unique sequence $(f_n)_{n\in\mathbb{N}}$ of elements of $H(D)$ such that $f_0 \in H(\widetilde{D})$, $f_n \in H_0(K\backslash T_n)$ whenever $n \in \mathbb{N}^*$, and such that the series $\sum\limits_{n=0}^{\infty} f_n$ converges to $f$ in $H(D)$.*

*Moreover we have*

$$\|f\|_D = \max\left(\|f_0\|_D, \sup_{n\in\mathbf{N}^*} \|f_n\|_D\right)$$

*and*

$$\|f_0\|_D = \|f_0\|_{\widetilde{D}}, \|f_n\|_D = \|f_n\|_{K\setminus T_n}\ (n \in \mathbf{N}^*).$$

Every one of the holes $T_n$ is called a *f-hole*, and $f_n$ is called *the Mittag-Leffler term of f associated to* $T_n$.

Now if $T$ is one of the $f$-holes, and $h$ is the Mittag-Leffler term of $f$ associated to $T$, we will just denote $h$ by $f_T$.

LEMMA 1. *Let D be an infraconnected closed bounded set, let* $T = d(a, r^-)$ *be a hole of D and let* $f \in H_0(K\setminus T)$. *Let* $b \in T$. *Then* $f(x)$ *expands into Laurent series* $\sum_{n=1}^{\infty} \frac{\lambda_n}{(x-a)^n} = \sum_{n=1}^{\infty} \frac{\mu_n}{(x-b)^n}$ *and then we have* $\lambda_1 = \mu_1$ *and* $|\lambda_1| \leq r\|f\|_D$.

Since the coefficient $\lambda_1$ does not depend on the point $a$ in $T$, we will call it *the Residue of f on the hole T*, and we will denote it by $R_T(f)$.
By definition we have $|R_T(f)| \leq r\|f\|_D$.

More generally for every $f \in H(D)$, $R_T(f_T)$ will be called *the Residue of f on the hole T* and we put $R_T(f) = R_T(f_T)$. Therefore by definition we have Relation $(\chi)$

$$(\chi) \qquad |R_T(f)| \leq r\|f_T\|_{K\setminus T} \leq r\|f\|_D.$$

*In the sequel, D will denote a clopen bounded infraconnected set.*

Let $f \in H(D)$. Then $f$ will be said to be *integrable* in $H(D)$ if there exists $F \in H(D)$ such that $F' = f$.

LEMMA 2. *Let D be a clopen bounded infraconnected set and let f be an integrable element in* $H(D)$. *For every hole T of D, we have* $R_T(f) = 0$.

Now Lemma 3 lets us introduce the quasi-integrable elements.

LEMMA 3. *Let D be a clopen bounded infraconnected set and let* $f \in H(D)$ *be such that* $R_T(f) = 0$ *for all the holes T of D, but finitely many of them* $T_1, \ldots, T_q$, *let* $\alpha_j$ *and* $\beta_j \in T_j$ *and let* $\lambda_j = R_{T_j}(f)$ $(1 \leq j \leq q)$.

*If* $f - \sum_{j=0}^{q} \frac{\lambda_j}{x-\alpha_j}$ *is integrable in* $H(D)$ *then so is* $f - \sum_{j=1}^{q} \frac{\lambda_j}{x-\beta_j}$.

**DEFINITION.** In such a case, $f$ will be said to be *quasi-integrable* in $H(D)$.

Let us recall the definition of a copiercing sequence given in [$S_4$]. An infinite sequence $(a_n, b_n)_{n\in\mathbb{N}^*}$ is said to be a *copiercing sequence associated to* $D$ if:

a) for every $n \in \mathbb{N}^*$,
either both $a_n$ and $b_n$ belong to a same hole $T_n$ of $D$ of diameter $r_n$,
or both $a_n$ and $b_n$ belong to $K\backslash\widetilde{D}$ and then we put $r_n = 1$,

b) $\lim_{n\to\infty} \frac{|a_n-b_n|}{r_n} = 0$.

In the same way, a finite sequence $(a_n, b_n)_{1\leq n\leq q}$ is said to be a *copiercing sequence associated to* $D$ if it just satisfies a).

By Lemmas 2 and 3 of [$S_4$] we have the following Lemma 4.

**LEMMA 4.** *Let* $D$ *be a clopen bounded infraconnected set and let* $(a_n, b_n)_{n\in\mathbb{N}^*}$ *be a copiercing sequence associated to* $D$.

*The sequence* $m \to \prod_{n=1}^{m} \left(\frac{x-a_n}{x-b_n}\right)$ *converges in* $H(D)$ *to a Meromorphic Product* $f(x)$ *which satisfies* $\|f-1\|_D < 1$. *Moreover* $f$ *is invertible in* $H(D)$.

We will denote by exp the exponential function $\sum_{n=0}^{\infty} \frac{x^n}{n!}$ whose convergence radius is 1 if the residue characteristic of $K$ is 0 and is $p^{-\frac{1}{p-1}}$ if the residue characteristic is $p \neq 0$ [A].
Henceforth we will denote by $R$ this convergence radius.

We will denote by log the logarithm function defined in $d(1, 1^-)$ by

$$\log(1+x) = \sum_{n=0}^{\infty}(-1)^n \frac{x^{n+1}}{n+1} \quad (x \in d(0, 1^-)). \qquad \text{[A]}$$

**THEOREM 1.** *Let* $D$ *be a clopen bounded infraconnected set, let* $\omega \in H(D)$ *and let* $r \in ]0, R[$. *The equation* $\ell(\omega)$ *has solutions* $g$ *invertible in* $H(D)$ *if and only if* $\omega$ *has the form*

$$\sum_{i=1}^{q} \frac{u_i}{x-c_i} + \sum_{n=1}^{t(r)} \left(\frac{1}{x-a_n} - \frac{1}{x-b_n}\right) + \frac{d\Omega}{dx}$$

*where:*

a) *each* $c_i$ *belongs to a hole* $T_i$ *with* $T_j \neq T_i$ *whenever* $j \neq i$, $u_i \in \mathbf{Z}$, $u_i = R_{T_i}(\omega)$

b) *the finite sequence* $(a_n, b_n)_{1 \leq n \leq t(r)}$ *is a copiercing sequence associated to* $D$, *and depends on* $r$

c) $\Omega \in H(D)$ *and* $\|\Omega\|_D \leq r$

d) *for every hole* $T$ *of* $D$ *other than* $T_1, \dots T_q$, $R_T(\omega) = 0$.

*Moreover if Conditions* a), b), c), d) *are satisfied, the solutions of* $\ell(\omega)$ *are of the form*

$$\lambda \left( \prod_{i=1}^{q} (x - c_i)^{u_i} \right) \left( \prod_{n=1}^{t(r)} \left( \frac{x - a_n}{x - b_n} \right) \right) \exp(\Omega(x)) \text{ with } \lambda \in K.$$

**DEFINITION.** A set $D$ will be said to be *well pierced* if there exists $\rho > 0$ such that every hole of $D$ has a diameter $r \geq \rho$.

**Remark.** A well pierced set is obviously open. When $D$ is well pierced Theorem 1 may be translated into the following Theorem 2.

**THEOREM 2.** *Let* $D$ *be a well pierced closed bounded infraconnected set and let* $\omega \in H(D)$. *The following conditions* a) *and* b) *are equivalent.*

a) $\omega$ *expands in the form*

$$\sum_{i=1}^{q} \frac{u_i}{x - c_i} + \sum_{n=1}^{\infty} \frac{1}{x - a_n} - \frac{1}{x - b_n}$$

*where* $u_i \in \mathbf{Z}^*$, $c_i$ *belongs to a hole* $T_i$ *of* $D$ *with* $T_j \neq T_i$ *whenever* $j \neq i$, *and where the sequence* $(a_n, b_n)_{n \in \mathbf{N}^*}$ *is a copiercing sequence associated to* $D$.

b) *The equation* $\ell(\omega)$ *admits solutions invertible in* $H(D)$.

*Moreover if these conditions are satisfied, then* $R_{T_i}(\omega) = u_i$ *whenever* $i = 1, \dots, q$, $R_T(\omega) = 0$ *for every hole* $T$ *different from* $T_1, \dots, T_q$ *and the solutions of* $\ell(\omega)$ *are the functions*

$$g(x) = \lambda \left( \prod_{i=1}^{q} (x - c_i)^{u_i} \right) \left( \prod_{n=1}^{\infty} \frac{x - a_n}{x - b_n} \right) \quad \text{with} \quad \lambda \in K.$$

We are now able to characterize the quasi-integrable elements (provided $D$ is well pierced).

**THEOREM 3.** *Let $D$ be a well pierced closed bounded infraconnected set and let $\omega \in H(D)$. Then $\omega$ is integrable (resp. quasi-integrable) in $H(D)$ if and only if $\omega$ is of the form*

$$\frac{1}{p^h}\sum_{n=1}^{\infty}\frac{1}{x-a_n}-\frac{1}{x-b_n}$$

$$(\text{resp. } \sum_{i=1}^{q}\frac{\lambda_i}{x-\alpha_i}+\frac{1}{p^h}\sum_{n=1}^{\infty}\frac{1}{x-a_n}-\frac{1}{x-b_n})$$

*with $h \in \mathbf{N}$, $(a_n, b_n)_{n\in\mathbf{N}^*}$ a copiercing sequence associated to $D$ (resp. and where each $\alpha_i$ belongs to a hole $T_i$ of $D$ with $T_i \neq T_j$ whenever $i \neq j$, and $\lambda_i = R_{T_i}(\omega)$).*

**COROLLARY.** *Let $D$ be a well pierced bounded infraconnected set. For every $f \in H(D)$ there exists an integer $h \in \mathbf{N}$ and a copiercing sequence $(a_n, b_n)_{n\in\mathbf{N}^*}$ associated to $D$ such that $f(x) = \frac{1}{p^h}\log\Pi(x)$ with $\Pi(x) = \prod_{n=1}^{\infty}\left(\frac{x-a_n}{x-b_n}\right)$.*

## 2. SOLUTIONS ANALYTIC FUNCTIONS IN $d(0,1^-)$

In all of this paragraph, $B$ will denote $d(0,1^-)$, so that the holes of $B$ are the disks $d(a,1^-)$ with $|a| = 1$. We will denote by $A(B)$ the algebra of the bounded power series convergent in $B$.

For every integer $q \geq 2$, the function "$\sqrt[q]{1+x}$" $= \sum_{n=0}^{\infty}\binom{\frac{1}{q}}{n}x^n$ which we will denote by $\phi_q(x)$ is known to have a convergence radius equal to 1 as long as the residue characteristic $p$ does not divide $q$, and $\frac{R}{p^m}$ when $q$ is of the form $p^m \cdot s$ with $s$ prime to $p$ [Ko].

**THEOREM 4.** *For every integer $q \geq 2$, $\phi_q$ does not belong to $H(B)$.*

**Remark 1.** Let $\omega \in H(B)$. Every residue $\lambda$ of $\omega$ satisfies $|\lambda| \leq \|\omega\|_B$ because the holes of $B$ have diameter 1. Moreover if $\ell(\omega)$ admits non zero solutions $f(x) \in A(B)$ then $|f'(x)| \leq |f(x)|$ whenever $x \in B$ (because $f$ has no zero in $B$ hence $|f(x)| = |f(0)|$ whenever $x \in B$). Then $\|\omega\|_B \leq 1$.

We will apply Theorem 1 and 2 by studying the functions $f \in A(B)$ which satisfy $\ell(\omega)$ while some power $f^N$ belong to $H(B)$. After an easy study when the residue characteristic $p$ is zero we will examine the case $p \neq 0$.

**Remark 2.** Let $\omega \in H(B)$ be such that $\ell(\omega)$ admits non zero solutions $f$. The set $J$ of the integers $n \in \mathbf{Z}$ such that $f^n \in H(B)$ is an ideal of $\mathbf{Z}$ which does not depend on the solution $f \neq 0$.

Indeed since $f$ is a solution of $\ell(\omega)$, $f$ has no zero in $B$, hence when $f^n \in H(B)$, $f^n$ is a unit in $H(B)$. Now it is easily seen that given $m, n \in J$, $q \in \mathbf{Z}$, then $f^{(m-n)q} = (f^m)^q\left((f^{-1})^n\right)^q$ does belong to $H(B)$, hence $(m-n)q \in J$.

Since the space of the solutions of $\ell(\omega)$ in $A(B)$ obviously has dimension 1, it does not depend on the solution $f$ so that we will denote it by $J(\omega)$.

$\alpha$) **We first suppose $p = 0$.**

Let $\phi \in A(B)$; we know that $\phi$ has integrals in $A(B)$ and we will denote by $\psi$ the one that satisfies $\psi(0) = 0$. Then we have $\|\psi\|_B = \|\phi\|_B$. The differential equation $y' = \phi y$ admits non zero solutions in $A(B)$ if and only if $\|\phi\|_B \leq 1$, and when $\|\phi\|_B \leq 1$, these solutions are the functions $\lambda \exp(\psi(X))$.

Thanks to Theorem 2 we can easily characterize the $f \in H(B)$ such that the equation $\ell(\omega)$ admits non zero solutions in $A(B)$ which satisfy $f^N \in H(B)$ for some $N \in \mathbf{N}^*$.

THEOREM 5. *Let $\omega \in H(B)$. The two following assertions are equivalent.*

i) *There exist non zero solutions $f$ of $\ell(\omega)$ in $A(B)$ and an integer $N \in \mathbf{N}^*$ such that $f^N \in H(B)$*

ii) *$N\omega$ is of the form*

$$\sum_{i=1}^{q} \frac{u_i}{x-c_i} + \sum_{n=1}^{\infty} \frac{1}{x-a_n} - \frac{1}{x-b_n}$$

*where each $c_i$ belongs to a hole $T_i$ of $B$ with $T_j \neq T_i$ whenever $j \neq i$, each $u_i$ belongs to $\mathbf{Z}^*$ and $u_i = R_{T_i}(\omega)$ and where the sequence $(a_n, b_n)$ is a copiercing sequence associated to $B$.*

**Remark.** Theorem 5 consists of saying that $J(\omega)$ is equal to the set of the integers $N$ satisfying ii).

$\beta$) **We now suppose $p \neq 0$.**

Let us recall that the convergence radius of the $p$-adic exponential is $R = p^{-\frac{1}{p-1}}$. Theorem 6 now is a direct application of Theorem 2.

**THEOREM 6.** *Let $\omega \in H(B)$ be such that $\ell(\omega)$ admits non zero solutions $f \in A(B)$ and let $N \in \mathbb{N}^*$. The following conditions* a) *and* b) *are equivalent*

a) $f^N \in H(B)$,

b) *$\omega$ is of the form*

$$\frac{1}{N}\left(\sum_{i=1}^{q}\frac{u_i}{x-c_i}+\sum_{n=1}^{\infty}\frac{1}{x-a_n}-\frac{1}{x-b_n}\right)$$

*where the sequence $(a_n, b_n)$ is a copiercing sequence associated to $B$ and where each $c_i$ belongs to a hole $T_i$ of $B$ with $T_j \neq T_i$ whenever $j \neq i$, $u_i \in \mathbb{Z}^*$, $\frac{u_i}{N} \in \mathbb{Q}^* \cap \mathbb{Z}_p$, $u_i = NR_{T_i}(\omega)$ and $R_T(\omega) = 0$ for every hole $T$ of $B$ different from $T_1, \dots, T_q$.*

In terms of quasi-integrability we obtain the following Theorem 7 which applies to the Frobenius Structure.

**DEFINITION.** Let $\omega \in H(B)$; $\omega$ will be said to have *finitely many residues, all of them in a set $A \subset K$*, if for every hole $T$ of $B$, $R_T(\omega) = 0$ except for finitely many holes $T$ such that $R_T(f) \in A$.

**THEOREM 7.** *Let $\omega$ be such that $\ell(\omega)$ has non zero solutions $f \in A(B)$.*

*There exists $N \in \mathbb{N}$ such that $f^N \in H(B)$ if and only if $\omega$ is quasi-integrable and has finitely many residues all of them in $\mathbb{Q}$.*

We will now look for a sufficient condition for $\ell(\omega)$ to admit non zero solutions $f \in A(B)$ such that a certain power $f^N$ belongs to $H(B)$.

**THEOREM 8.** *Let $\omega$ belong to $A(B)$ and be in the form $\sum_{i=1}^{s} \frac{\lambda_i}{x-\alpha_i} + \frac{d\Omega}{dx}$ with $\lambda_i \in \mathbb{Q}^* \cap \mathbb{Z}_p$, $|\alpha_i| \geq 1$ $(q \leq i \leq s)$ and $\Omega \in A(B)$, $\|\Omega\|_B \leq R$.*

*Then $\ell(\omega)$ admits non zero solutions $f \in A(B)$. Moreover if $\Omega \in H(B)$, then for every $N \in \mathbb{N}$ such that $\|N\Omega\|_B < R$ and $N\lambda_i \in \mathbb{Z}$ $(1 \leq i \leq s)$ we have $f^N \in H(B)$.*

The following conjecture is now suggested by Theorem 8 and by the case when $\omega \in K$ with $|\omega| = R$.

**Conjecture**

*Let $\omega \in H(B)$. The equation $\ell(\omega)$ admits non zero solutions $f \in A(B)$ such that $f^N \in H(B)$ for some $N \in \mathbb{N}$ if and only if $\omega$ is of the form*

$$\sum_{j=1}^{s} \frac{u_j}{x-c_j} + \frac{d\Omega}{dx} \text{ with } u_j \in \mathbb{Q} \cap \mathbb{Z}_p \ (1 \le i \le s), \text{ and } |c_j| = 1, \Omega \in H(B)$$

*and $\|\Omega\|_B < R$.*

The very simple example $\omega = \frac{1}{p}\left(\frac{1}{x-a} - \frac{1}{x-b}\right)$ helps us understand the problem.

If $|a-b| \le \frac{R}{p}$, the function $\log\left(\frac{x-a}{x-b}\right)$ belongs to $H(B)$ and satisfies $\left|\log\left(\frac{a-a}{x-b}\right)\right| < |a-b|$ for all $x \in B$ hence $\left|\frac{1}{p}\log\left(\frac{x-a}{x-b}\right)\right| < p|a-b| \le R$ and then the function $f(x) = \exp\left(\frac{1}{p}\log\left(\frac{x-a}{x-b}\right)\right)$ belongs to $A(B)$ so that $f$ is a solution of $\ell(\omega)$ which satisfies $f^p \in H(B)$.

Now if $\frac{R}{p} < |a-b| \le R$, $\ell(p\omega)$ still admits $g(x) = \frac{b}{a}\left(\frac{x-a}{x-b}\right)$ for solution.

But $\ell(\omega)$ can't have a non zero solution in $A(B)$. Indeed suppose $\ell(\omega)$ has a non zero solution in $A(B)$.

Then $f^p(x) = g(x) = \frac{b}{a}\left(\frac{x-a}{x-b}\right) = \frac{b}{a}\left(1 + \frac{b-a}{x-a}\right)$ and then (by the classical properties of the Taylor series in B) we know that $g(B) = d(1, |b-a|^-)$. Moreover, without loss of generality we can assume $f(0) = 1$ because $g(0) = 1$.

Then $f(B)$ is a disk $d(1, r^-)$ and it is easily seen that $r > R$. Indeed we know that $|(1+u)^p - 1| = 1/_p|u|$ whenever $|u| < R$ hence if $r \le R$ then $|g(x) - 1| \le R/_p$ whenever $x \in B$ hence $|b-a| \le R/_p$ which contradicts the hypothesis, hence $r > R$.

But then we know that the function $u \to (1+u)^p$ is not injective in $d(0, r^-)$ if $r > R$, hence the function $(f(x))^p$ is not injective in $B$, while $g$ is. That finishes proving that $\ell(\omega)$ has no solution $f \in A(B)$ but zero.

Let us now recall the problem of the existence of a Frobenius Structure on a differential equation $\ell(\omega)$.

Let us assume that $\omega \in H(B)$, $\omega$ has finitely many residues and $\ell(\omega)$ admits a solution $f \in A(B)$. Then $\ell(\omega)$ is said *to have a Frobenius Structure* if there exists $h \in \mathbb{N}$ such that $\frac{f(x)}{f(x^{(p^h)})} \in H(B)$ [$C_1$,R,D].

It was already proven that $\ell(\omega)$ has Frobenius Structure provided $f$ is algebraic over the quotients field of $H(B)$ [$C_2$]. In particular, this holds when $f^N \in H(B)$ for some $N \in \mathbb{N}$. Then we have an obvious corollary of Theorem 7.

COROLLARY. *Let $\omega$ be quasi-integrable in $H(B)$, with finitely many residues, all of them in $\mathbb{Q}$ and such that $\ell(\omega)$ admits non zero solutions $f \in A(B)$.*

*Then $\ell(\omega)$ has a Frobenius Structure.*

Actually the previous Corollary generalizes the existence of a Frobenius Structure obtained when $f$ extends to a superadmissible set $D$, [D1] as the following Theorem 9 shows.

THEOREM 9. *Let $D$ be a set of the form $d(0,S)\backslash\left(\bigcup_{i\in I} d(\alpha_i,r)\right)$ with $|\alpha_i-\alpha_j| = |\alpha_i| = 1$ whenever $i \neq j$ and $r < 1 < S$ and let $\omega \in H(D)$ have finitely many non zero residues. There exists a set $D' = d(0,S')\backslash\left(\bigcup_{i\in I} d(\alpha_i,r'^-)\right)$ with $r < r' < 1 < S' < S$ such that $\omega$ is quasi-integrable in $H(D')$.*

**Remark.** It is easily seen that Theorem 9 has no reciprocal. We can easily construct an element $\omega$ quasi-integrable in $H(B)$ which has no extension to any set of this kind.

## 3. PROOFS

*Proof of Lemma* 1. In $H(K\backslash T)$, $\frac{1}{x-b}$ is of the form $\frac{1}{x-a}\sum_{m=0}^{\infty}\left(\frac{b-a}{x-a}\right)^m$, so that for every $n \in \mathbb{N}^*$ $\frac{1}{(x-b)^n}$ has no term in the form $\frac{\lambda}{x-a}$ and thereby the only term in $\frac{1}{x-a}$ comes from $\frac{\mu_1}{x-b}$, i.e. $\frac{\mu_1}{x-a}\sum_{m=0}^{\infty}\left(\frac{b-a}{x-a}\right)^m$. Thus, we see that $\mu_1 = \lambda_1$.
Moreover by Mittag-Leffler's Theorem [K,R] we know that

$$\|f\|_D = \|f\|_{K\backslash T} = \sup_{n\in N}\frac{|\lambda_n|}{r^n}$$

hence we see that $|\lambda_1| \leq r\|f\|_D$.

*Proof of Lemma* 2. Let $F \in H(D)$ be such that $F' = f$. Let $T$ be a hole of $D$. First let us assume $F$ to belong to $H_0(K\backslash T)$. Then $F$ is of the form $\sum_{n=1}^{\infty}\frac{\lambda_n}{(x-a)^n}$ while $f = F' = -\sum_{n=1}^{\infty}\frac{n\lambda_n}{(x-a)^{n+1}}$ hence $R_T(f) = 0$ and of course $R_\tau(f) = 0$ for every hole $\tau$ other than $T$ because $f \in H_0(K\backslash T)$.

Now let us consider the general case when $f$ and $F \in H(D)$, and let $\sum_{n=0}^{\infty} F_n$ be the Mittag-Leffler series of $F$ with $F_0 \in H(\widetilde{D})$, and for each $n \in \mathbb{N}^*$, $F_n \in H_0(K\backslash T_n)$.

By Theorem 3 of [E5] we know that the Mittag-Leffler series of $f$ is just $\sum_{n=0}^{\infty} F'_n$. Then by the result already shown when $f \in H_0(K\backslash T)$ we see that for every hole $T$ of $D$ we have $R_T(F'_n) = 0$ whenever $n \geq 1$, (and of obviously $R_T(F'_0) = 0$) hence $R_T(f) = 0$.

*Proof of Lemma* 3. It is easily seen that we may assume $q = 1$ with no loss of generality and then we put $\lambda = \lambda_1$, $\alpha = \alpha_1$, $\beta = \beta_1$, and $T = T_1$.

$$f - \frac{\lambda}{x-\beta} = f - \frac{\lambda}{x-\alpha} + \lambda\left(\frac{1}{x-\alpha} - \frac{1}{x-\beta}\right)$$

Hence we just have to check that $\frac{1}{x-\alpha} - \frac{1}{x-\beta}$ has an integral in $H(D)$. Let $r$ be the diameter of $T$; then $|\alpha-\beta| < r$ hence $\log\left(1 + \frac{\beta-\alpha}{x-\beta}\right) \in H_0(K\backslash T)$ (because $\left|\frac{\beta-\alpha}{x-\beta}\right| \leq \frac{|\beta-\alpha|}{r}$ whenever $x \in K\backslash T$). Thus $\log\left(\frac{x-\alpha}{x-\beta}\right)$ belongs to $H(D)$ and is an integral of $\frac{1}{x-\alpha} - \frac{1}{x-\beta}$, and that ends the proof of Lemma 3.

Let us recall that Lemma 4 is given by Lemma 2 and 3 of [S4].

The classical Lemma 5 may be found in [E2].

LEMMA 5. *Let $D$ and $D'$ be two sets in $K$, let $f \in H(D)$ be such that $f(D) \subset D'$ and let $g \in H(D')$. Then $g \circ f$ belongs to $H(D)$.*

We will often use the following application.

COROLLARY. *Let $D$ be a closed bounded set in $K$, let $f \in H(D)$ be such that $\|f-1\|_D < 1$ and let $g \in H(D)$ be such that $\|g\|_D < R$. Then the functions $\log(f(x))$ and $\exp(g(x))$ belong to $H(D)$.*

PROPOSITION P. *Let $D$ be a clopen bounded infraconnected set, let $(a_n, b_n)_{n\in\mathbb{N}^*}$ be a copiercing sequence associated to $D$ and let $f$ be the Meromorphic Product $\prod_{n=1}^{\infty}\left(\frac{x-a_n}{x-b_n}\right) \in H(D)$. We assume that $f' \in H(D)$. Let $\omega = \frac{f'}{f}$.*

*For every $r > 0$, there exists $t(r) \in \mathbb{N}$ and $\Omega \in H(D)$ such that $\|\Omega\|_D < r$, $\frac{d\Omega}{dx} \in H(D)$ and $\omega = \sum_{n=1}^{t(r)}\left(\frac{1}{x-a_n} - \frac{1}{x-b_n}\right) + \frac{d\Omega}{dx}$.*

*Proof.* Let $I_0 = \{n \in \mathbb{N}$ such that $b_n \in \widetilde{D}\}$ and let $(T_m)_{m\in\mathbb{N}^*}$ be the sequence of the holes that contain at least one term of the sequence $(b_n)$.

For every $m \in \mathbb{N}^*$, let $I_m = \{n \in \mathbb{N} \mid b_n \in T_m\}$. By Theorem 3 of [S4] we know that $f$ factorizes in the form $\prod_{m=0}^{\infty} f_m$ with $f_m(x) = \prod_{n\in I_m}\left(\frac{x-a_n}{x-b_n}\right)$ and $\lim_{m\to\infty} \|f_m-1\|_D = 0$.

Moreover, by Lemma 3 we have $\|f_m - 1\|_D < 1$ so that $\log(f_m)$ belongs to $H(D)$ whenever $n \in \mathbb{N}$, and the series $\sum_{m\in\mathbb{N}} \log(f_m)$ converges in $H(D)$.

More precisely, $\log(f_0) \in H(\widetilde{D})$ and $\log(f_m) \in H(K\backslash T_m)$ $(m \in \mathbb{N}^*)$, so that the series $\sum_{m=0}^{\infty} \log f_m$ is just the Mittag-Leffler series of $\log f$ on the infraconnected set $D$ [R].

Since $f' \in H(D)$ obviously $\frac{d}{dx}(\log f) = \frac{f'}{f} \in H(D)$ and then by Theorem 3 of [E5] we know that the Mittag-Leffler series of $\frac{d}{dx}(\log f)$ is just $\sum_{m=0}^{\infty} \frac{f'_m}{f_m}$.

Now since the sequence $(a_n, b_n)_{n\in\mathbb{N}^*}$ is copiercing, there exists $t(r) \in \mathbb{N}$ such that $\|\frac{x-a_n}{x-b_n} - 1\|_D < \min(r, R)$ for every $n > t(r)$ and then we know that $\|\log\left(\frac{x-a_n}{x-b_n}\right)\|_D < r$. Let $\Omega = \sum_{n=t(r)+1}^{\infty} \log\left(\frac{x-a_n}{x-b_n}\right)$.

We have $\|\Omega\|_D < r$ and $\omega = \sum_{n=1}^{t(r)} \left(\frac{1}{x-a_n} - \frac{1}{x-b_n}\right) + \frac{d\Omega}{dx}$.

That ends the proof of Proposition P.

**LEMMA 6.** *Let $T$ be a hole of a bounded closed infraconnected set $D$ and let $a, b \in T$. Then for every hole $\tau$ of $D$ we have $R_\tau\left(\frac{1}{x-a} - \frac{1}{x-b}\right) = 0$.*

*Proof.*

If $\tau = T$, $R_T\left(\frac{1}{x-a}\right) = 1 = R_T\left(\frac{1}{x-b}\right)$ hence $R_\tau\left(\frac{1}{x-a} - \frac{1}{x-b}\right) = 0$.

If $\tau \neq T$, $R_\tau\left(\frac{1}{x-a}\right) = R_\tau\left(\frac{1}{x-b}\right) = 0$, hence $R_\tau\left(\frac{1}{x-a} - \frac{1}{x-b}\right) = 0$.

**LEMMA 7.** *Let $D$ be a closed bounded infraconnected set and let $(a_n, b_n)$ be a copiercing sequence associated to $D$ such that the series $\sum_{n=1}^{\infty} \frac{1}{x-a_n} - \frac{1}{x-b_n}$ converges in $H(D)$ and let $\omega$ be its sum. For every hole $T$ of $D$ we have $R_T(\omega) = 0$.*

*Proof.* Relation $(\chi)$ shown in paragraph 1 (after Lemma 1) shows that $R_T$ is continuous on $H(D)$ and Lemma 6 gives $R_T\left(\frac{1}{x-a_n} - \frac{1}{x-b_n}\right) = 0$ hence $R_T(\omega) = 0$.

*Proof of Theorem* 1. We first suppose that a) b) c) d) are satisfied. Then $\exp(\Omega(x))$ belongs to $H(D)$ and it is easily seen that the functions $\lambda \prod_{i=1}^{q} (x-a)^{u_i} \prod_{n=1}^{t(r)} \left(\frac{x-a_n}{x-b_n}\right)$ $\exp(\Omega(x))$ (with $\lambda \in K$) are solutions of $\ell(\omega)$. They are invertible elements of $H(D)$ and then by Theorem 1 of [E6] the space of solutions of $\ell(\omega)$ has dimension 1, hence all the solutions have this form.

Conversely, we assume $\ell(\omega)$ to have a solution $g$ invertible in $H(D)$. By Theorem 2 of [S5], $g$ factorizes as $F(x)h(x)$ with $F \in R(D)$, $F$ of the form $\prod_{i=1}^{q}(x-c_i)^{u_i}$, each $c_i$ belongs to a hole $T_i$ of $D$ and $T_j \neq T_i$ whenever $j \neq i$, $u_i$ belongs to $\mathbb{Z}^*$, and with $h$ a Meromorphic Product $\prod_{n=1}^{\infty}\left(\frac{x-a_n}{x-b_n}\right)$ where $(a_n, b_n)_{n\in\mathbb{N}^*}$ is a copiercing sequence associated to $D$. Then $\omega = \frac{F'}{F} + \frac{h'}{h}$.

By Proposition P, there exists $t(r) \in \mathbb{N}$ and $\Omega \in H(D)$ such that $\|\Omega\|_D < r$, $\frac{d\Omega}{dx} \in H(D)$ and such that $\frac{h'}{h} = \sum_{n=1}^{t(r)}\left(\frac{1}{x-a_n} - \frac{1}{x-b_n}\right) + \frac{d\Omega}{dx}$.

Let us put $\phi = \sum_{n=1}^{t(r)}\left(\frac{1}{x-a_n} - \frac{1}{x-b_n}\right) + \frac{d\Omega}{dx}$; by Lemma 7 we see that $R_T(\phi) = 0$ for every hole $T$ of $D$, $R_{T_i}\left(\frac{F'}{F}\right) = u_i$ for every $i = 1, \ldots, q$ and $R_T\left(\frac{F'}{F}\right) = 0$ for every hole $T$ different from $T_1, \ldots, T_q$.

We see that Conditions b) and c) are already satisfied and it only remains to show that $R_T\left(\frac{d\Omega}{dx}\right) = 0$ for every hole $T$, to conclude that a) and d) are satisfied too.

Let $T$ be a hole, let $\rho$ be its diameter.

Let $I(T)$ be the set of the $n > t(r)$ such that $a_n$ and $b_n$ belong to $T$. The series $\sum_{n\in I(T)} \frac{1}{x-a_n} - \frac{1}{x-b_n}$ converges to a limit $\psi$ and by Lemma 7 we see that (1) $R_T(\psi) = 0$.

On the other hand it is easily seen that $\psi = \frac{d}{dx}\left(\sum_{n\in I(T)} \log\left(\frac{x-a_n}{x-b_n}\right)\right)$ because $\left\|\frac{d}{dx}\left(\log\left(\frac{x-a_n}{x-b_n}\right)\right)\right\|_D = \left\|\frac{1}{x-a_n} - \frac{1}{x-b_n}\right\|_D = \frac{|a_n-b_n|}{\rho}$ whenever $n \in I(T)$.

Moreover when $n > t(r)$ and $n \notin I(T)$, $\log\left(\frac{x-a_n}{x-b_n}\right)$ does belong to $H(D\cup T)$ hence $\sum_{n>t(r)} \log\left(\frac{x-a_n}{x-b_n}\right) \in H(D\cup T)$ and thereby, $\frac{d\Omega}{dx} - \psi \in H(D\cup T)$ hence $R_T\left(\frac{d\Omega}{dx} - \psi\right) = 0$.

Finally by (1) we have $R_T\left(\frac{d\Omega}{dx}\right) = 0$. This is true for every hole $T$ of $D$ and then Conditions a) and d) are now proven, and Theorem 1 is proven too.

**Remark.** In the general case when $D$ is not well pierced we can't assert that the series $\sum_{n=1}^{\infty} \frac{1}{x-a_n} - \frac{1}{x-b_n}$ converges in $H(D)$ because the convergence of the series is not uniform in $D$ in the general case. This is the reason why we have to consider separately $\psi$ and $\frac{d\Omega}{dx} - \psi$.

When $D$ is well pierced the situation is more simple as shows Lemma 8.

**LEMMA 8.** *Let $D$ be a well pierced closed bounded infraconnected set and $(a_n, b_n)_{n\in\mathbb{N}^*}$ be a copiercing sequence associated to $D$.*
*Let $h$ be a Meromorphic Product $\prod_{n=1}^{\infty}\left(\frac{x-a_n}{x-b_n}\right)$.*
*Then $h' \in H(D)$ and the series $\sum_{n=1}^{\infty} \frac{1}{x-a_n} - \frac{1}{x-b_n}$ converges to $\frac{h'}{h}$ in $H(D)$.*

*Proof.* Since $D$ is well pierced let $\delta > 0$ be such that every hole of $D$ has diameter $\rho \geq \delta$. By Theorem 1 of [E5] we know that for every $f \in H(D)$, $f'$ belongs to $H(D)$ and $\|f'\|_D \leq \frac{1}{\delta}\|f\|_D$.

For every $q \in \mathbb{N}^*$ let $h_q = \prod_{n=1}^{q} \frac{x-a_n}{x-b_n}$; then $\frac{h'_q}{h_q} = \sum_{n=1}^{q} \frac{1}{x-a_n} - \frac{1}{x-b_n}$.

Now by Lemma 4 we have $\|h_q - 1\|_D < 1$ and $\|h-1\|_D < 1$ hence $|h(x)| = |h_q(x)| = 1$ for all $x \in D$, and then

$$\left|\frac{h'_q(x)}{h_q(x)} - \frac{h'(x)}{h(x)}\right| = |h'_q(x)h(x) - h'(x)h_q(x)| \leq$$
$$\leq \max(|h'_q(x) - h'(x)|\ |h(x)|, |h'(x)|\ |h(x) - h_q(x)|).$$

But $|h(x)| = 1$, $\|h'\|_D \leq \frac{1}{\delta}\|h\|_D = \frac{1}{\delta}$ and $\|h'_q - h'\|_D \leq \frac{1}{\delta}\|h_q - h\|_D$ so that it is easily seen that $\left|\frac{h'_q(x)}{h_q(x)} - \frac{h'(x)}{h(x)}\right| \leq \frac{1}{\delta}\|h_q - h\|_D$ whenever $x \in D$ hence $\left\|\frac{h'_q}{h_q} - \frac{h'}{h}\right\|_D \leq \frac{1}{\delta}\|h_q - h\|_D$. Then the sequence $q \to \frac{h'_q}{h_q}$ converges to $\frac{h'}{h}$ and that ends the proof of Lemma 8.

*Proof of Theorem 2.* We first assume a) to be satisfied. Since the sequence $(a_n, b_n)_{n\in\mathbb{N}^*}$ is a copiercing sequence associated to $D$, the Meromorphic Product $h(x) = \prod_{n=1}^{\infty}\left(\frac{x-a_n}{x-b_n}\right)$ obviously converges in $H(D)$ by Lemma 1 of [S4]. Let us put $F(x) = \prod_{i=1}^{q}(x-c_i)^{u_i}$ and $g(x) = F(x)h(x)$; then $g \in H(D)$ and $\frac{g'}{g}(x) = \omega(x)$ whenever $x \in D$ hence $h$ is a solution of $\ell(\omega)$. Obviously, $F$ and $h$ are invertible in $H(D)$ hence so is $g$, and then the space of the solutions has dimension 1 by Theorem 1 of [E6]. Hence if a) is satisfied, the solutions of $\ell(\omega)$ are the functions

$$\lambda\left(\prod_{i=1}^{q}(x-c_i)^{u_i}\right)\left(\prod_{n=1}^{\infty}\frac{x-a_n}{x-b_n}\right) \quad \text{with } \lambda \in K.$$

Then we know that $R_{T_j}\left(\sum_{i=1}^{q}\frac{u_i}{x-c_i}\right) = u_j$ and $R_T\left(\sum_{i=1}^{q}\frac{u_i}{x-c_i}\right) = 0$ for every hole $T$ different from $T_1, \ldots T_q$ and by Lemma 7 we have $R_T\left(\sum_{n=1}^{\infty}\left(\frac{1}{x-a_n} - \frac{1}{x-b_n}\right)\right) = 0$ for every hole $T$ of $D$.

So we have proven that a) implies b) and the additional properties.

Conversely, let us assume $\ell(\omega)$ to have an invertible solution $g \in H(D)$, and let us show that a) satisfied.

By Corollary of [S$_4$] $g$ factorizes in the form $F(x)h(x)$ where $h$ is a Meromorphic Product $\prod_{n=1}^{\infty}\left(\frac{x-a_n}{x-b_n}\right)$ with $(a_n, b_n)_{n\in\mathbb{N}^*}$ a copiercing sequence associated to $D$, and $F$ is a rational function in the form $\prod_{i=1}^{q}(x-c_i)^{u_i}$, with $u_i \in \mathbb{Z}^*$, each $c_i$ belongs to a hole $T_i$ of $D$ such that $T_j \neq T_i$ whenever $j \neq i$.

By Lemma 7, $\frac{h'}{h}$ is equal to $\sum_{n=1}^{\infty} \frac{1}{x-a_n} - \frac{1}{x-b_n}$ hence $\omega = \sum_{i=1}^{q} \frac{u_i}{x-c_i} + \sum_{n=1}^{\infty} \frac{1}{x-a_n} - \frac{1}{x-b_n}$, and that finishes proving Theorem 2.

*Proof of Theorem* 3. We first suppose $\omega$ to be of the form $\eta(x) = \sum_{n=1}^{\infty} \frac{1}{x-a_n} - \frac{1}{x-b_n}$ with $(a_n, b_n)$ a copiercing sequence associated to $D$) and let us show that $\eta$ is integrable in $H(D)$. Let $f$ be the Meromorphic Product $\prod_{n=1}^{\infty}\left(\frac{x-a_n}{x-b_n}\right)$. Then $\eta = \frac{f'}{f}$ and by Proposition P, for $r < 0$, $\eta$ has the form $\sum_{n=1}^{t(r)} \frac{1}{x-a_n} - \frac{1}{x-b_n} + \frac{d\Omega}{dx}$ with $\Omega \in H(D)$ (and $\|\Omega\|_D < r$). By the Corollary of Lemma 5 we know that $\sum_{n=1}^{t(r)} \log \frac{x-a_n}{x-b_n} \in H(D)$ hence $\eta$ clearly appears to be the derivative of an element of $H(D)$ so that $\omega$ is integrable.

More generally if $\omega$ has the form $\sum_{i=1}^{q} \frac{\lambda_i}{x-\alpha_i} + \eta(x)$ with each $\alpha_i$ in a hole $T_i$ such that $T_i \neq T_j$ whenever $i \neq j$, and with $\eta$ as above, then $\omega$ is clearly quasi-integrable in $H(D)$.

Now let us suppose $\omega$ to be quasi-integrable, and let $\omega = \sum_{i=1}^{q} \frac{\lambda_i}{x-\alpha_i} + \frac{d\Omega}{dx}$ with $\Omega \in H(D)$. Let $h \in \mathbb{N}$ be such that $\|p^h\Omega\|_D < R$. The equation $\ell\left(p^h \frac{d\Omega}{dx}\right)$ clearly has solutions $f \in H(D)$, in the form $\lambda \exp(p^h\Omega)$, and then $p^h \frac{d\Omega}{dx}$ has the form announced in Theorem 1, but by hypothesis $\frac{d\Omega}{dx}$ only has zero residues hence $p^h \frac{d\Omega}{dx}$ is in the form $\sum_{n=1}^{\infty} \frac{1}{x-a_n} - \frac{1}{x-b_n}$ with $(a_n, b_n)$ a copiercing sequence associated to $D$. Thus a quasi-integrable $\omega$ has the form announced in Theorem 3. Finally, if $\omega$ is integrable in $H(D)$ then by Lemma 4, $\omega$ only has zero residues hence $\lambda_i = 0$ whenever $i = 1, \dots, q$ and then $\omega$ is just in the form $\sum_{n=1}^{\infty} \frac{1}{x-a_n} - \frac{1}{x-b_n}$. That finishes proving Theorem 3.

The following Lemma 9 will be helpful in Theorems 4, 5, 6, 7.

**LEMMA 9.** *Let $\omega \in H(B)$ and let $S$ (resp. $T$) be the space the solutions of $\ell(\omega)$ in $H(B)$ (resp. $A(B)$). If $S \neq \{0\}$ then $S = T$.*

*Proof.* We know that $T$ has dimension 0 or 1. If $S \neq \{0\}$ then $S$ has dimension 1 and then $S = T$.

*Proof of Theorem 4.* $\phi_q$ obviously satisfies the equation $y' = \frac{1}{n(1+x)}y$ defined in $B$. Suppose $\phi_q \in H(B)$. Let $T = d(-1, 1^-)$; then $R_T\left(\frac{1}{n(1+x)}\right) = \frac{1}{n}$ which contradicts the fact that $R_T\left(\frac{1}{n(1+x)}\right) \in \mathbb{Z}$ (see Theorem 2). Hence $\phi_q \notin H(B)$.

*Proof of Theorem 5.* We first suppose $N$ to belong to $J(\omega)$. Then by Theorem 2 $N\omega$ is in the form announced in ii).

Now let us assume $N$ to be in the form announced in ii).

By Theorem 2, $\ell(N\omega)$ admits invertible solutions $g$ in $H(D)$, hence $|g(x)|$ is constant in $B$.

Let $h = \frac{g}{g(0)}$; since $h(0) = 1$ it is easily seen that $|h(x)-1| < 1$ whenever $x \in B$, hence the function $f$ defined in $B$ by $f(x) = \phi_N(h(x))$ belongs to $A(B)$. Then $f$ is a solution of $\ell(\Omega)$, and then $f^N$ is a solution of $\ell(N\omega)$. By Theorem 2, $\ell(N\omega)$ has non zero solutions in $H(B)$. Hence by Lemma 9 every solution in $A(B)$ belongs to $H(B)$, which finishes the proof of Theorem 5.

*Proof of Theorem 6.* We first suppose a) is satisfied. Let $g = f^N$. Then $g$ is a solution of $\ell(N\omega)$ in $H(B)$. Hence $N\omega$ is in the form a) of Theorem 2: $\sum_{i=1}^{q} \frac{u_i}{x-c_i} + \sum_{n=1}^{\infty} \frac{1}{x-a_n} - \frac{1}{x-b_n}$ hence $\omega$ is clearly in the form b) of Theorem 6.

Now we suppose b) is satisfied. Let $S$ (resp. $T$) be the linear space of the solutions of $\ell(N\omega)$ in $H(B)$ (resp. in $A(B)$). Since $\ell(\omega)$ admits a non zero solution $f$ in $A(B)$, obviously $f^N$ is a solution of $\ell(N\omega)$ in $A(B)$. Now by Theorem 2, $\ell(N\omega)$ has non zero solutions in $H(B)$. Hence by Lemma 9, $f^N$ belongs to $H(B)$ and that ends the proof of Theorem 6.

*Proof of Theorem 7.* If $\omega$ has a non zero solution $f$ such that $f^N \in H(B)$, by Theorem 2 and Theorem 3 we know that $N\omega$ is quasi-integrable and by Theorem 6, $\omega$ has finitely many residues all of them in $\mathbb{Q} \cap \mathbb{Z}_p$.

Now let us suppose $\omega$ is quasi-integrable and has finitely many residues, all of them in $\mathbb{Q}$.

Since $\ell(\omega)$ has non zero solutions $f \in A(B)$ we know that the residue $\lambda_i$ of $\omega$ satisfy $|\lambda_i| \leq 1$ hence $\lambda_i \in \mathbf{Q} \cap \mathbf{Z}_p$.

Then by Theorem 3, $\omega$ is of the form b) of Theorem 6, hence by Theorem 6 there exists an integer $N$ such that $f^N \in H(B)$ and that ends proof of Theorem 7.

*Proof of Theorem* 8. By considering $\phi = \Omega - \Omega(0)$ which satisfies the same hypothesis as $\Omega$ we can assume $\Omega(0) = 0$ with no loss of generality. Then by Lemma 6 we know that $|\Omega(x)| < R$ whenever $x \in B$. Hence $\exp(\Omega(x))$ belongs to $A(B)$ and is a solution of the equation $y' = \left(\frac{d\Omega}{dx}\right) y$.

Now let $q$ be the smallest common multiple of the denominators of the $\lambda_i$ $(1 \leq i \leq s)$ and let $v_i = q\lambda_i \in \mathbf{Z}$ $(1 \leq i \leq s)$. We see that $\prod_{i=1}^{s}(x-\alpha_i)^{v_i}$ belongs to $R(B)$ and is a solution of the equation $y' = \left(\sum_{i=1}^{s} \frac{v_i}{x-\alpha_i}\right) y$.

Therefore the function $h(x) = \prod_{i=1}^{s} \left(\frac{\alpha_i - x}{\alpha_i}\right)^{v_i} \exp(q\Omega(x))$ belongs to $A(B)$ and is a solution of $\ell(q\omega)$.

Now since $|\alpha_i| \geq 1$ whenever $i = 1, \ldots, s$ we have $|g(x) - 1| < 1$ for all $x \in B$. On the other hand, since the $\lambda_i$ belong to $\mathbf{Z}_p$, $q$ is prime to $p$ hence the function $g$ defined in $B$ by $g(x) = \phi_q(h(x))$ belongs to $A(B)$ and is a solution of $\ell(\omega)$.

Finally if $\Omega \in H(B)$ and if $\|Ng\|_B < R$ then we know that $\exp(N(\Omega) \in H(B)$; obviously $\left(\frac{\alpha_1 - x}{\alpha_1}\right) \in R(B)$ $(1 \leq i \leq s)$, so that $g^N$ belongs to $H(B)$ and that ends the proof of Theorem 8.

LEMMA 10. *Let $r, r'$ be such that $0 < r < r'$. Let $T = d(0, r^-)$. Let $T' = d(0, r'^-)$ and let $f \in H_0(K \backslash T)$ be such that $R_T(f) = 0$.*

*The sequence $n \to \left(\frac{r}{r'}\right)^n \frac{1}{|n-1|}$ $(n \geq 2)$ has limit zero and there exists $F \in H_0(K \backslash T')$ such that the restriction of $f$ to $K \backslash T'$ is equal to $F'$ in $H_0(K \backslash T')$ and $\|F\|_{K \backslash K'} \leq \|f\| \sup_{n \geq 2} \frac{1}{|n-1|} \left(\frac{r'}{r}\right)^n$.*

*Proof.* Since $|n-1| \geq \frac{1}{n-1}$ the sequence $\left(\frac{r}{r'}\right)^n$ obviously has limit 0.
Let $A = \sup_{n \geq 2} \left(\frac{r}{r'}\right)^n \frac{1}{|n-1|}$. Since $f \in H_0(K \backslash T)$ and since $R_T(f) = 0$, $f(x)$ is a Laurent series in the form $\sum_{n=2}^{\infty} \frac{\lambda_n}{x^n}$ with $\lim_{n \to \infty} \frac{|\lambda_n|}{r^n} = 0$. Let $\mu_{n-1} = \frac{\lambda_n}{n-1}$; then $\frac{|\mu_{n-1}|}{r'^n} \leq rA\left(\frac{|\lambda_n|}{r^n}\right)$ hence $\lim_{n \to \infty} \frac{|\mu_n|}{r'^{n-1}} = 0$ and the series $F(x) = \sum_{n=1}^{\infty} \frac{\mu_n}{x^n}$ belongs to $H_0(K \backslash T')$ and $F'(x) = f(x)$ whenever $x \in K \backslash T$.

*Proof of Theorem* 9. With no loss of generality, we may obviously assume $\omega$ to have all its residues equal to zero.

Let $A = \sup_{n\geq 2} \frac{1}{|n-1|}\left(\frac{r}{r'}\right)^{n-1}$, for each $i \in I$. Let $T_i = d(\alpha_i, r^-)$ and let $T_i' = d(\alpha_i, r')$ and let $\omega = f_0 + \sum_{i\in I} f_i$ with $f_0 \in H(D)$ and $f_i \in H_0(K\backslash T_i)$.

By Lemma 10, for each $i \in I$ there exists $F_i \in H_0(K\backslash T_i)$ such that $F_i'(x) = f_i(x)$ whenever $x \in K\backslash T_i'$ and $\|F_i\|_{K\backslash T_i'} = \|F_i\|_{D'} \leq Ar\|f_i\|_D$ so that the series $\sum_{i\in I} F_i$ converges in $H(D')$.

In the same way, as $\lim_{n\to\infty}\left(\frac{S'}{S}\right)^n \frac{1}{|n-1|} = 0$, the power series $f_0(x) = \sum_{n=0}^{\infty} \alpha_n x^n$ admits an integral in the form $F_0(x) = \sum_{n=0}^{\infty} \frac{\alpha_n}{n+1} x^{n+1}$ which converges in $d(0, S') = \widetilde{D}$. Let us put $\Omega = F_0 + \sum_{i\in I} F_i$; then by Theorem 3 of [E5] we know that $\Omega' = \omega$ and that ends the proof of Theorem 9.

## REFERENCES


[A] Amice, Y.: *Les nombres p-adiques.* P.U.F. (1975).

[C1] Christol, G.: *Solutions algébriques des équations différentielles p-adique.* Progr. Math. **38** (1983), 51–58.

[C2] Christol, G.: *Fonctions et éléments algébriques.* Pacific J. Math. **125** (1986), 1–37.

[D1] Dwork, B. and Robba, P.: *On ordinary linear p-adic differential equations.* Trans. Amer. Math. Soc. **231** (1977), 1–46.

[D2] Dwork, B.: *Lectures on p-adic Differential Equations.* Springer-Verlag, New York-Heidelberg-Berlin, 1982.

[E1] Escassut, A.: *Les algèbres de Krasner.* C.R. Acad. Sci. Paris Sér. A-B **272** (1971), A598–A601.

[E2] Escassut, A.: *Algèbres d'éléments analytiques en analyse non archimédienne.* Nederl. Akad. Wetensch. Proc. Ser. A **77** = Indag. Math. **36** (1974), 339–251.

[E3] Escassut, A.: *Eléments analytiques et filtres percés sur un ensemble infraconnexe.* Ann. Mat. Pura Appl. (4) **110** (1976), 335–352.

[E4] Escassut, A.: *T-filtres, ensembles analytiques et transformation de Fourier p-adique.* Ann. Inst. Fourier (Grenoble) **25** (1975), 45–80.

[$E_5$] Escassut, A.: *Derivative of Analytic Elements on infraconnected clopen sets.* Nederl. Akad. Wetensch. Indag. Math. **51** (1989), 63–70.

[$E_6$] Escassut, A. and Sarmant, M.C.: *The differential equation $y' = fy$ in the algebras $H(D)$.* Collect. Math. **39** (1988), 31–40.

[$E_7$] Escassut, A. and Sarmant, M.C.: *The p-adic Differential Equation $y' = \omega y$ in the closed unit disk.* To appear in Collectanea Mathematica (Barcelona).

[Ko] Koblitz, N.: *p-Adic Numbers, p-Adic Analysis and Zeta-Function.* Springer-Verlag, New York-Heidelberg-Berlin, 1977.

[K] Krasner, M.: *Prolongement analytique uniforme et multiforme dans les corps valués complets.* In *Les tendances géometriques en algèbre et théorie des nombres.* Clermont-Ferrand 1964, p.97–141. Centre National de la Recherche Scientifique (1966) (Colloques Internationaux du C.N.R.S. Paris, 143).

[M] Motzkin, E.: *La décomposition d'un élément analytique en facteurs singuliers.* Ann. Inst. Fourier (Grenoble) **27** (1977), 67–82.

[R] Robba, Ph.: *Fonctions analytiques sur les corps valués ultramétriques complets, Prolongement analytique et algèbres de Banach ultramétriques.* Astérisque, **10** (1973), 109–220.

[$S_1$] Sarmant, M.C., Escassut, A.: *T-suites idempotentes.* Bull. Sci. Math. (2) **106** (1982), 289–303.

[$S_2$] Sarmant, M.C.: *Produits Méromorphes.* Bull. Sci. Math. **109** (1985) 155–178.

[$S_3$] Sarmant, M.C., Escassut, A.: *Fonctions Analytiques et Produits Croulants.* Collect. Math. **36** (1985), 199–218.

[$S_4$] Sarmant, M.C., Escassut, A.: *Prolongement analytique à travers un T-filtre.* Studia Sci. Math. Hungar. **22** (1987), 407–444.

[$S_5$] Sarmant, M.C.: *Factorisation en Produit Méromorphe d'un élément semi-inversible.* To appear in the Bulletin des Sciences Mathématiques, Paris.

Université Pierre et Marie Curie, Mathématiques
Tour 45-46, 5ème étage
4 Place Jussieu
F-75230 Paris 05, France

Université Blaise Pascal
Mathématiques Pures
F-63177 Aubiere Cedex, France

# THE $p$-ADIC KREIN-ŠMULIAN THEOREM

W.H. Schikhof

**ABSTRACT.** The natural non-archimedean version of the Krein-Šmulian Theorem holds essentially only when the base field is spherically complete (Corollary 1.6). For Banach spaces over nonspherically complete scalar fields two restricted versions of the Krein-Šmulian Theorem (Theorems 2.2 and 3.1) are proved.

**INTRODUCTION.** (For unexplained terms see below.) Consider the following statement (*).

$$(*) \qquad \begin{cases} \textit{Let } E \textit{ be a } K\textit{-Banach space and let } A \subset E' \textit{ be convex. If } A \cap B \\ \textit{is } w'\textit{-closed in } B \textit{ for each bounded set } B \subset E' \textit{ then } A \textit{ is } w'\textit{-closed.} \end{cases}$$

If $K = \mathbb{R}$ or $\mathbb{C}$ then (*) is known as the Krein-Šmulian Theorem. By modifying classical techniques (in particular by using $c$-compactness arguments) (*) can also be proved if $K$ is a spherically complete non-archimedean valued field ([4], Theorem 5.1). Now see the Abstract.

**PRELIMINARIES.** (For terms still unexplained see [2].) Throughout $K$ is a nonarchimedean nontrivially valued field that is complete under the metric induced by the valuation $|\ \ |$. We always ASSUME THAT $K$ IS NOT SPHERICALLY COMPLETE, so $|K| := \{|\lambda| : \lambda \in K\}$ is dense. If $X \subset K$ is bounded, nonempty, we set diam $X = \sup\{|x - y| : x, y \in X\}$.

Let $E$ be a $K$-vector space. A nonempty subset $A$ of $E$ is *absolutely convex* if $x, y \in A$, $\lambda, \mu \in K$, $|\lambda| \le 1$, $|\mu| \le 1$ implies $\lambda x + \mu y \in A$. For such $A$ we set $A^e := \bigcap\{\lambda A : \lambda \in K,\ |\lambda| > 1\}$. $A$ is *edged* if $A = A^e$. The smallest absolutely convex set containing $X \subset E$ is denoted co $X$. A nonempty set in $E$ is *convex* (*edged convex*) if it is an additive coset of an absolutely convex (edged absolutely convex) set. By definition, the empty set is convex. The *algebraic dual* of $E$ is the vector space $E^*$ consisting of all linear functions $E \to K$. The weakest topology on $E$ for which all $f \in E^*$ are continuous is denoted $\sigma(E, E^*)$.

A *seminorm* on $E$ is a map $p : E \to [0, \infty)$ such that $p(x) \geq 0$, $p(\lambda x) = |\lambda| p(x)$, $p(x+y) \leq \max(p(x), p(y))$ for all $x, y \in E$, $\lambda \in K$. We shall use expressions such as '$p$-convergence', '$p$-closure', '$p$-compactoid', '$p$-orthogonal' without further explanation. A seminorm $p$ is *of finite type* if Ker $p$ has finite codimension, *of countable type* if $E/\text{Ker } p$ with the norm induced by $p$ is of countable type. A seminorm $p$ is *polar* if $p = \sup\{|f| : f \in E^*,\ |f| \leq p\}$. A seminorm $p$ is a *norm* if $p(x) = 0$ implies $x = 0$. Norms are usually denoted $\|\ \ \|$ rather than $p$.

Let $(E, \|\ \ \|)$ be a normed space over $K$. Let $a \in E$, $r > 0$. We write $B_E(a, r) := \{x \in E : \|x-a\| \leq r\}$ and $B_E := B_E(0, 1)$. The *dual space* $E'$ is the Banach space consisting of all continuous linear functions $E \to K$, normed by $f \mapsto \|f\| := \sup_{B_E} |f|$. The natural map $j_E : E \to E''$ is continuous. $E$ is *pseudoreflexive* if $j_E$ is an isometry (which is equivalent to polarity of the norm on $E$). A linear map $T$ from a $K$-Banach space $E$ to a $K$-Banach space $F$ is a *quotient map* if $T$ maps $\{x \in E : \|x\| < 1\}$ onto $\{x \in F,\ \|x\| < 1\}$.

Let $(E, \tau)$ be a locally convex space over $K$. It is called *of finite* (*countable*) *type* if every continuous seminorm is of finite (countable) type. $(E, \tau)$ is *strongly polar* if each continuous seminorm is polar, *polar* if there exists a base of polar continuous seminorms. Let $E' = (E, \tau)'$ be the space of all continuous linear functions $E \to K$. The *weak topology* $w = \sigma(E, E')$ is the weakest topology on $E$ such that all $f \in E'$ are continuous. Similarly, the *weak-star topology* $w' = \sigma(E', E)$ is the weakest topology on $E'$ such that for each $x \in E$ the evaluation $f \mapsto f(x)$ $(f \in E')$ is continuous. It is well known (see [5]) that the natural map $E \to (E', \sigma(E', E))'$ is surjective. Let $X \subset E$, $Y \subset E'$. We set $X^0 := \{f \in E' : |f(x)| \leq 1 \text{ for all } x \in X\}$ and $Y_0 := \{x \in E : |f(x)| \leq 1 \text{ for all } f \in Y\}$. $X$ is a *polar set* if $X^0_0 = X$. For a ball $B_E(0, r)$ in a normed space $E$ we have $B_E(0, r)^0 = B_{E'}(0, 1/r)$. If $E$ is pseudoreflexive, $B_{E'}(0, r)_0 = B_E(0, 1/r)$. The closure of a set $X \subset E$ is $\overline{X}$. Instead of $\overline{\text{co } X}$ we write $\overline{\text{co}}\, X$. Let $E, F$ be locally convex spaces over $K$. The *adjoint* of a continuous linear map $T : E \to F$ is the map $T' : F' \to E'$ defined by $f \mapsto f \circ T$. Following [1] we say that a subspace $D$ of $E$ has the Weak Extension Property (WEP) if the adjoint

$E' \to D'$ of the inclusion map $D \hookrightarrow E$ is surjective.

## §1. FAILURE OF THE KREIN-ŠMULIAN THEOREM

The key theorem of this section is the following. Recall that $K$ is not spherically complete.

**THEOREM 1.1.** *Let $\tau_1, \tau_2$ be locally convex topologies on a $K$-vector space $E$ such that $\tau_2$ is of finite type while $\tau_1$ is not. Then there exists a $\tau_1$-closed absolutely convex set in $E$ that is not $\tau_2$-closed.*

We signal the following corollary which is in sharp contrast to the theory over spherically complete base fields.

**COROLLARY 1.2.** *Let $E$ be a locally $K$-convex space whose topology is not the weak topology. Then there exists a closed absolutely convex set in $E$ that is not weakly closed.*

The proof of Theorem 1.1 runs in a few steps. Let us say that a seminorm $q$ on a $K$-vector space is *special* if $q(x) \in |K|$ for each $x \in E$ and if for all $x, y \in E$

$$x \perp y \text{ in the sense of } q \quad \Longrightarrow \quad q(x) = 0 \text{ or } q(y) = 0$$

**LEMMA 1.3.** *On a normed space of countable type over $K$ there exists an equivalent special norm.*

*Proof.* Let $(\check{K}, |\ \ |)$ be the spherical completion of $(K, |\ \ |)$ in the sense of [2], Theorem 4.49. Then $|\ \ |$, considered as a norm on the $K$-vector space $\check{K}$ is special. (Indeed, we have $|\check{K}| = |K|$. If $x, y \in \check{K}$, $x \perp y$, $y \neq 0$ then $xy^{-1} \perp 1$ so $xy^{-1} \perp K$. But $\check{K}$ is an immediate extension of $K$ so $xy^{-1} = 0$ i.e. $x = 0$.) As $\check{K}$ is infinite dimensional over $K$ we can, for a given normed space $E$ of countable type over $K$, make a $K$-linear homeomorphism $T$ of $E$ into $\check{K}$. Then $x \mapsto |Tx|$ is the required norm.

**LEMMA 1.4.** *Let $E$ be a strongly polar locally convex space over $K$. If $E$ is not of finite type then there exists a continuous special seminorm $q$ on $E$, $q$ not of finite type.*

*Proof.* There is a continuous seminorm of infinite type $p$ on $E$. The $p$-continuous linear functions form an infinite dimensional space so we can find linearly independent

$f_1, f_2, \ldots \in E'$ such that $n|f_n| \leq p$ for each $n \in \mathbb{N}$. The formula $\widetilde{p}(x) = \max_n |f_n(x)|$ defines a continuous seminorm $\widetilde{p}$ on $E$, of infinite countable type. Now Lemma 1.3 (applied to $E/\text{Ker } \widetilde{p}$) leads to a special seminorm $q$ equivalent to $\widetilde{p}$.

**Remark.** The conclusion of Lemma 1.4 holds for any polar space $E$ that is not of finite type.

LEMMA **1.5.** *Let $q$ be a special seminorm on a $K$-vector space $E$. If $q$ is not of finite type then $\{x \in E : q(x) < 1\}$ is not $\sigma(E, E^*)$-closed.*

*Proof.* Let $x \in E$, $q(x) = 1$ (such $x$ exist!). We shall prove that $x$ is in the $\sigma(E, E^*)$-closure of $A := \{x \in E : q(x) < 1\}$ by producing, for given $f_1, \ldots, f_n \in E^*$, a point $a \in A$ such that $f_i(x-a) = 0$ for $i \in \{1, \ldots, n\}$.

(i) Suppose $f_1(x-a) \neq 0$ for all $a \in A$. Then $f_1(x) \notin f_1(A)$ so, by convexity, $f_1(A)$ is bounded and $f_1$ is $q$-continuous. We have

$$|f_1(x)| \geq \text{diam } f_1(A) = \sup\{|f_1(a)| : q(a) < 1\} = \|f_1\|$$

where $\|f_1\|$ is the operator seminorm of $f_1$ with respect to $q$. For each $y \in \text{Ker } f_1$

$$\|f_1\| q(x-y) \geq |f_1(x-y)| = |f_1(x)| \geq \|f_1\|$$

and we find $x \perp y$ in the sense of $q$. As $q$ is special and $q(x) = 1$ we must have $q = 0$ on $\text{Ker } f_1$ implying that $q$ is of finite type, a contradiction. Thus, we may conclude that there exists an $a_1 \in A$ with $f_1(x-a_1) = 0$.

(ii) Now we repeat the argument in (i) where $E$ is replaced by $\text{Ker } f_1$, $q$ by $q|\text{Ker } f_1$, $x$ by $x-a_1$, $A$ by $A \cap \text{Ker } f_1$ and $f_1$ by $f_2|\text{Ker } f_1$. (Indeed, $q|\text{Ker } f_1$ is special, of infinite type and $q(x-a_1) = 1$). So there exists an $a_2 \in A \cap \text{Ker } f_1$ such that $f_2(x-a_1-a_2) = 0$. Observe that also $f_1(x-a_1-a_2) = 0$. In this spirit we arrive inductively at points $a_1, a_2, \ldots, a_n \in A$ such that $f_i(x-a) = 0$ ($i \in \{1, \ldots, n\}$) where $a := \sum_{i=1}^{n} a_i \in A$.

*Proof of Theorem* 1.1. If $(E, \tau_1)$ is not strongly polar, choose any nonpolar continuous seminorm $q$ and set $A := \{x \in E : q(x) \leq 1\}$. $A$ is $\tau_1$-closed but, as $q$ is not polar and $A$ is edged, $A$ is not $\sigma(E, E^*)$- closed so certainly $A$ is not $\tau_2$-closed. If $(E, \tau_1)$ is strongly polar, let $q$ be as in Lemma 1.4. By Lemma 1.5 the set $A := \{x \in E : q(x) < 1\}$ is not $\sigma(E, E^*)$-closed, so not $\tau_2$-closed.

Part (i) of the next corollary demonstrates the failure of the Krein-Šmulian Theorem for nonspherically complete base fields.

COROLLARY 1.6. *Let $E$ be a normed space over $K$ such that $E'$ is infinite dimensional.*

(i) *There exists an absolutely convex set $A \subset E'$ such that $A \cap B$ is $w'$-closed in $B$ for each bounded set $B \subset E'$ while $A$ is not $w'$-closed.*

(ii) *There exists an absolutely convex set $A \subset E$ such that $A \cap B$ is $w$-closed in $B$ for each bounded set $B \subset E$ while $A$ is not $w$-closed.*

*Proof.* (i) $(E', w')$ is an infinite dimensional Hausdorff space of countable type so its dual (which is $j_E(E)$) is infinite dimensional. Thus, we can choose $x_1, x_2, \ldots$ in $E$ such that $j_E(x_1), j_E(x_2), \ldots$ are linearly independent and $\lim_{n\to\infty} \|x_n\| = 0$. The seminorm $p$ on $E'$ defined by

$$p(f) = \max_n |f(x_n)| = \max_n |j_E(x_n)(f)|$$

is therefore not of finite type. By Theorem 1.1 there exists an absolutely convex set $A \subset E'$ which is $p$-closed but not $w'$-closed. But it is easily seen that, on any bounded set $B \subset E'$, $w'$-convergence implies $p$-convergence. Thus, the $p$-closedness of $A$ implies that $A \cap B$ is $w'$-closed in $B$.

(ii) Similar to the above proof but now with the seminorm $x \mapsto \max_n |f_n(x)|$ $(x \in E)$, where $f_1, f_2, \ldots$ is a linearly independent sequence in $E'$ for which $\lim_{n\to\infty} \|f_n\| = 0$. We leave the details to the reader.

## §2. SAVE THE KREIN-ŠMULIAN THEOREM! (PART ONE)

To save the Krein-Šmulian Theorem we shall concentrate on *edged* convex sets. As such sets are translates of edged absolutely convex sets no harm is done by considering only the latter. Thus, we arrive at

DEFINITION 2.1. A normed space $E$ over $K$ is a *Krein-Šmulian space* if the following holds. If $A \subset E'$ is absolutely convex and edged and if $A \cap B$ is $w'$-closed in $B$ for each bounded set $B \subset E'$ then $A$ is $w'$-closed.

Observe that, for an absolutely convex $A \subset E'$, the expression '$A \cap B$ is $w'$-closed in $B$ for each bounded set $B \subset E'$' is equivalent to 'for each $n \in \mathbb{N}$ the set $A \cap B_{E'}(0, n)$

is $w'$-closed' and, if $A$ is a subspace, to '$A \cap B_{E'}$ is $w'$-closed'.

The main result of this section is

**THEOREM 2.2.** *A strongly polar Banach space is a Krein-Šmulian space.*

For the proof we need first a lemma on Banach spaces. Let us call a sequence $X_1 \supset X_2 \supset \ldots$ of closed absolutely convex subsets of a $K$-Banach space $E$ *quasi Cauchy* if for each $\lambda \in K$, $|\lambda| > 1$ and $N \in \mathbf{N}$

$$X_n \subset \lambda(X_m + B_E(0, \frac{1}{N})) \qquad (m, n \geq N)$$

**LEMMA 2.3.** *Let $X_1 \supset X_2 \supset \ldots$ be a quasi Cauchy sequence in a $K$-Banach space. Set $X := \bigcap_n X_n$. Then, for each $n \in \mathbf{N}$ and $x_n \in X_n$, and each $\lambda \in K$, $|\lambda| > 1$ there is an $x \in \lambda X$ such that $\|x_n - x\| \leq \frac{|\lambda|}{n}$.*

*Proof.* Choose $\lambda_1, \lambda_2, \ldots \in K$ with $|\lambda_i| > 1$ for each $i$, $\prod_i |\lambda_i| = |\lambda|$. We have

$$X_n \subset \lambda_1(X_{n+1} + B_E(0, \frac{1}{n})) \quad \text{whence} \quad X_n \subset \lambda_1 X_{n+1} + B_E(0, \frac{|\lambda|}{n}))$$

$$X_{n+1} \subset \lambda_2(X_{n+1} + B(0, \frac{1}{n+1}) \quad \text{whence} \quad \lambda_1 X_{n+1} \subset \lambda_1\lambda_2 X_{n+2} + B_E(0, \frac{|\lambda|}{n+1})$$

etc.

So, given $x_n \in X_n$, we can find a sequence $x_{n+1}, x_{n+2}, \ldots$ where $x_{n+1} \in \lambda_1 X_{n+1}$, $x_{n+2} \in \lambda_1\lambda_2 X_{n+2}, \ldots$ such that for all $k \in \{0, 1, 2, \ldots\}$

$$\|x_{n+k} - x_{n+k+1}\| \leq \frac{|\lambda|}{n+k}.$$

By completeness $x := \lim_{k\to\infty} x_{n+k}$ exists. We have $\lambda^{-1}x_{n+1} \in \lambda^{-1}\lambda_1 X_{n+1} \subset X_{n+1}$; $\lambda^{-1}x_{n+2} \in \lambda^{-1}\lambda_1\lambda_2 X_{n+2} \subset X_{n+2}$, etc., so $\lambda^{-1}x = \lim_{k\to\infty} \lambda^{-1}x_{n+k} \in \bigcap_{i \geq n+1} X_i = X$ and it follows that $x \in \lambda X$. Further, we have

$$\|x_n - x\| \leq \max\{\|x_n - x_{n+1}\|, \|x_{n+1} - x_{n+2}\|, \ldots\} \leq \max(\frac{|\lambda|}{n}, \frac{|\lambda|}{n+1}, \ldots) \leq \frac{|\lambda|}{n}.$$

*Proof of Theorem 2.2.* Let $A \subset E'$ be absolutely convex, edged and assume that $A \cap B_{E'}(0, n)$ is $w'$-closed for each $n \in \mathbf{N}$. Then ($A \cap B_{E'}(0, n)$ is also edged) $A \cap B_{E'}(0, n)$ is a polar set. Setting

$$X_n := \left(A \cap B_{E'}(0, n)\right)_0 \qquad (n \in \mathbf{N})$$

$$X := \bigcap_n X_n$$

one verifies immediately (i), (ii), (iii), (iv) below.

(i) Each $X_n$ is a polar subset of $E$.

(ii) $X_n^0 = A \cap B_{E'}(0,n)$ for each $n \in \mathbb{N}$.

(iii) $X_1 \supset X_2 \supset \ldots$.

(iv) $X = A_0$.

(v) For each $N \in \mathbb{N}$ and $m, n \geq N$

$$X_n \subset (X_m + B_E(0, \frac{1}{N}))_0^0.$$

(Proof: $\left(X_m + B_E(0, \frac{1}{N})\right)^0 = X_m^0 \cap B_E(0, \frac{1}{N})^0 = A \cap B_{E'}(0,m) \cap B_{E'}(0,N) = A \cap B_{E'}(0,N)$, so $X_n \subset X_N = \left(A \cap B_{E'}(0,N)\right)_0 = \left(X_m + B_E(0, \frac{1}{N})\right)_0^0$.)

(vi) $X_1, X_2, \ldots$ is quasi Cauchy. (Proof. Let $\lambda \in K$, $|\lambda| > 1$, $N \in \mathbb{N}$, $m, n \geq N$. The set $X_m + B_E(0, \frac{1}{N})$ is norm open hence norm closed. So $\left(X_m + B_E(0, \frac{1}{N})\right)^e$ is norm closed and edged, hence polar (as $E$ is strongly polar). It follows via (v), that $X_n \subset \left(X_m + B_E(0, \frac{1}{N})\right)_0^0 \subset \lambda\left(X_m + B_E(0, \frac{1}{N})\right)$.)

(vii) $X^0 \subset A$. (Proof. Let $f \in X^0$, $\lambda \in K$, $|\lambda| > 1$. It suffices to prove that $f \in \lambda A$. Let $n \in \mathbb{N}$ be such that $\|f\| \leq n$. Choose any $x \in X_n$. By Lemma 2.3 there is a $y \in \lambda X$ with $\|x - y\| \leq \frac{|\lambda|}{n}$. We have

$$\begin{aligned}|f(x)| &\leq |f(x-y)| \vee |f(y)| \\ &\leq \|f\|\,\|x-y\| \vee |\lambda| \leq n \cdot \frac{|\lambda|}{n} \vee |\lambda| = |\lambda|\end{aligned}$$

and we see that $|\lambda^{-1} f| \leq 1$ on $X_n$, so $\lambda^{-1} f \in X_n^0 = A \cap B_{E'}(0,n) \subset A$ i.e. $f \in \lambda A$.)

Now (iv) combined with (vii) yields $A = X^0$ is $w'$-closed.

**COROLLARY 2.4.** *A subspace of the dual of a strongly polar Banach space is $w'$-closed as soon as its intersection with the closed unit ball is $w'$-closed.*

**COROLLARY 2.5.** *An edged absolutely convex subset $A$ of $\ell^\infty$ is $\sigma(\ell^\infty, c_0)$-closed as soon as $A \cap B_{\ell^\infty}(0,n)$ is $\sigma(\ell^\infty, c_0)$-closed for each $n \in \mathbb{N}$.*

*Proof.* $c_0$ is a (reflexive) strongly polar space.

We also have:

**THEOREM 2.6.** *If $E$ is a Krein-Šmulian space and $D \subset E$ is a closed subspace then $E/D$ is a Krein-Šmulian space.*

*Proof.* Let $i : (E/D)' \to E'$ be the adjoint of the quotient map $E \to E/D$. It is easily seen that $i$ is an isometry, that Im $i$ is $w'$-closed in $E'$ and that $i$ is a $w'$ to $w'$ homeomorphism $(E/D)' \to \text{Im } i$.

Now let $A$ be an edged absolutely convex subset of $(E/D)'$ such that $A \cap B$ is $w'$-closed in $B$ for each bounded set $B$ in $(E/D)'$. Then $i(A)$ is edged. If $X \subset E'$ is bounded then $i(A) \cap X$ is $w'$-closed in $X$. (Proof. Let $j \mapsto a_j$ be a net in $A$ such that $i(a_j) \in X$ for all $j$ and let $w' - \lim_j i(a_j) = b \in X$. As Im $i$ is $w'$-closed $b = i(a)$ for some $a \in i^{-1}(X)$. Then $w' - \lim_j a_j = a$. Now $a_j \in A \cap i^{-1}(X)$ for all $j$, $a \in i^{-1}(X)$ and $i^{-1}(X)$ is bounded, so by assumption on $A$ we have $a \in A$, so $b \in i(A) \cap X$.) Since $E$ is a Krein-Šmulian space, $i(A)$ is $w'$-closed in $E'$ so that $A = i^{-1}\big(i(A)\big)$ is $w'$-closed in $(E/D)'$.

**THEOREM 2.7.** *If $E$ is a Krein-Šmulian space and if $D \subset E$ is a weakly closed subspace having the WEP then $D$ is a Krein-Šmulian space.*

*Proof.* Let $\pi : E' \to D'$ be the adjoint of the inclusion map $D \hookrightarrow E$. Then $\pi$ is surjective and $w'$ to $w'$ continuous. If $A$ is an edged absolutely convex set in $D'$ and $\pi^{-1}(A)$ is $w'$-closed then $A$ is $w'$-closed. (Proof. Let $g \in D'$, $g \notin A$. There is an $f \in E'$ with $\pi(f) = g$. Then $f \notin \pi^{-1}(A)$. Now $\pi^{-1}(A)$ is $w'$-closed and edged so there exists an $x \in E$ such that $f(x) = 1$ and $|h(x)| < 1$ for all $h \in \pi^{-1}(A)$. In particular, $|h(x)| < 1$ for all $h \in \text{Ker } \pi = D^0$ i.e. $h(x) = 0$ for all $h \in D^0$ so $x \in D^0_0 = D$. Then $g(x) = f(x) = 1$ and $|h(x)| < 1$ for all $h \in A$.)

Now let $A$ be an absolutely convex edged subset of $D'$ such that $A \cap B$ is $\sigma(D', D)$-closed in $B$ for each bounded set $B \subset D'$. Then for such $B$, $\pi^{-1}(A) \cap \pi^{-1}(B)$ is $\sigma(E', E)$-closed in $\pi^{-1}(B)$. If $X \subset E'$ is bounded then $\pi(X)$ is bounded and $X \subset \pi^{-1}\big(\pi(X)\big)$ so it follows that $\pi^{-1}(A) \cap X$ is $w'$-closed in $X$ for each bounded set $X \subset E'$. Since $E$ is Krein-Šmulian we have that $\pi^{-1}(A)$ is $w'$-closed, so by the remark above, $A$ is $w'$-closed.

**Remark.** Not every Krein-Šmulian polar space is strongly polar; $\ell^\infty$ is an easy example. In §3 we will see that, if $I$ is large enough, $c_0(I)$ is not Krein-Šmulian. This leads to the

**PROBLEM.** Characterize the class of Krein-Šmulian spaces.

A concrete help would be the answer to the following two questions.

- Is $c_0 \times \ell^\infty$ a Krein-Šmulian space? (More generally, if $E_1$ and $E_2$ are Krein-Šmulian spaces then does it follow that $E_1 \times E_2$ is Krein-Šmulian?)
- Is the subspace of $D$ of $\ell^\infty$ constructed in [2], Ex. 4.J Krein-Šmulian?

## §3. SAVE THE KREIN-ŠMULIAN THEOREM! (PART TWO)

In this section we shall prove the following version of the Krein-Šmulian Theorem. Observe that ($\alpha$) holds for any polar $K$-Banach space.

**THEOREM 3.1.** *For a normed space $E$ over $K$ the following are equivalent.*
($\alpha$) *$j_E(E)$ is norm closed in $E''$.*
($\beta$) *If $H \subset E'$ is a subspace of finite codimension and if $H \cap B_{E'}$ is $w'$-closed then so is $H$.*

For a normed space $E$ over $K$ the $bw'$-topology (the 'bounded-weak-star topology') is by definition the strongest locally convex topology on $E'$ that coincides with $w'$ on bounded subsets of $E'$.

**PROPOSITION 3.2.** *Let $E$ be a normed space over $K$.*
(i) *$bw'$ is stronger than $w'$ but weaker than the norm topology on $E'$.*
(ii) *$(E', bw')$ is of countable type.*
(iii) *A seminorm $p$ on $E'$ is $bw'$-continuous if and only if $p|B_{E'}$ is $w'$-continuous.*
(iv) *For any locally convex space $(X, \tau)$ and any linear map $T : E' \to X$ we have that $T$ is $bw'$ to $\tau$ continuous if and only if $T|B'_E$ is $w'$ to $\tau$ continuous.*

*Proof.* $E' = [B_{E'}]$ and $B_{E'}$ is a $w'$-compactoid, hence a $bw'$-compactoid. This implies (ii). The other proofs are straightforward.

We know that $(E', w')' = j_E(E)$ ([5]). We now prove

**PROPOSITION 3.3.** *For a normed space $E$ over $K$ the dual of $(E', bw')$ is the norm closure of $j_E(E)$ in $E''$.*

*Proof.* Every $\theta \in \overline{j_E(E)}$ is, on $B_{E'}$, the uniform limit of a sequence in $j_E(E)$ so $\theta|B_{E'}$ is $w'$-continuous and $\theta$ is $bw'$-continuous (Proposition 3.2 (iv)). Thus $\overline{j_E(E)} \subset (E', bw')'$. Conversely, let $\theta \in (E', bw')'$. Then (Proposition 3.2 (i)) $\theta \in E''$. Let $\varepsilon > 0$; we shall find an $x \in E$ such that $\|\theta - j_E(x)\| < \varepsilon$. Let $\alpha \in K$, $0 < |\alpha| < \varepsilon$. The $w'$-continuity of $\theta | B_{E'}$ yields a finite set $F \subset E$ such that $f \in F^0 \cap B_{E'}$ implies $|\theta(f)| \leq |\alpha|$, in other words

$$f \in j_E(F)_0 \cap (B_{E''})_0 \quad \Longrightarrow \quad |(\alpha^{-1}\theta)(f)| \leq 1.$$

So we see that $\alpha^{-1}\theta \in \left(j_E(F)_0 \cap (B_{E''})_0\right)^0 = (A + B_{E''})_0^0$, where $A = j_E(\text{co } F)$. Now $B_{E''}$ is $w'$-closed and $A$ is finite dimensional so by [3], 1.4, $(A + B_{E''})_0^0 = (A + B_{E''})^e$. For any $\lambda \in K$ such that $|\lambda| > 1$ and $|\lambda\alpha| < \varepsilon$ we have $\alpha^{-1}\theta \in \lambda A + \lambda B_{E''}$ hence $\theta \in j_E(E) + \alpha\lambda B_{E''}$ and there is an $x \in E$ with $\theta - j_E(x) \in \alpha\lambda B_{E''}$ i.e. $\|\theta - j_E(x)\| < \varepsilon$.

**COROLLARY 3.4.** *Let $j_E(E)$ be closed in $E''$, let $A \subset E'$ be absolutely convex and edged. Then $A$ is $w'$-closed if and only if $A$ is $bw'$-closed.*

*Proof.* Let $A$ be $bw'$ closed. As $A$ is also edged and $(E', bw')$ is strongly polar (Proposition 3.2 (ii)), $A$ is a polar set i.e. $A = S_0$ for some $S \subset (E', bw')'$. But by Proposition 3.3 $S \subset (E', w')'$ so that $A$ is $w'$-closed.

Further, we need the following general lemma.

**LEMMA 3.5.** *Let $A$ be a closed absolutely convex subset of a Hausdorff locally convex space over $K$; let $D$ be a finite dimensional subspace such that $A \cap D = \{0\}$. Then $A + D$ is closed and the addition map is a homeomorphism $A \times D \to A + D$.*

*Proof.* (i) If addition is homeomorphic then $A + D$ is closed. In fact, let $i \to a_i + d_i$ be a net in $A + D$ (where $a_i \in A$, $d_i \in D$ for each $i$), converging to some $z$. Then $(i, j) \mapsto a_i - a_j + d_i - d_j$ converges to 0. By homeomorphism, $d_i - d_j \to 0$, by completeness of $D$, $d_i \to d$ for some $d \in D$. Then $a_i \to z - d$ and, by closedness of $A$, $z - d \in A$. We see that $z \in A + D$.

(ii) Assume $n := \dim D = 1$, say $D = Kx$ for some nonzero $x$. Let $i \mapsto a_i + \lambda_i x$ ($a_i \in A$, $\lambda_i \in K$) be a net in $A + D$ converging to 0. If not $\lambda_i \to 0$ we may assume $|\lambda_i| \geq |\alpha| > 0$ for all $i$ and some $\alpha \in K$. Then $\alpha\lambda_i^{-1}(a_i + \lambda_i x) \to 0$ so $\alpha x = -\lim_i \alpha\lambda_i^{-1} a_i \in \overline{A} = A$ conflicting $Kx \cap A = \{0\}$. Thus, addition is homeomorphic and via (i) the lemma is proved if $n = 1$.

(iii) The proof of the induction step $n-1 \to n$ is now standard and left to the reader.

*Proof of Theorem* 3.1.

(i) Suppose ($\alpha$), and let $H \subset E'$ be a subspace of finite codimension such that $H \cap B_{E'}$ is $w'$-closed. Then $H \cap B_{E'}$ is norm closed, hence so is $H$. For some $t \in (0, 1)$ $H$ has a

$t$-orthogonal complement $D$. Let $P : E' \to D$ be the obvious projection. For $\lambda \in K$, $|\lambda| \geq t^{-1}$ we have

$$B_{E'} \subset \lambda(H \cap B_{E'}) + \lambda(D \cap B_{E'}) \subset \lambda(H \cap B_{E'}) + D.$$

Let $i \mapsto f_i$ be a net in $B_{E'}$, $w' - \lim_i f_i = 0$. Then, by Lemma 3.5, $\lim_i Pf_i = 0$. We see that $P|B_{E'}$ is continuous, so (Proposition 3.2 (iv)) $P$ is $bw'$ to norm continuous and Ker $P = H$ is $bw'$-closed, hence $w'$-closed by Corollary 3.4, and $(\beta)$ is proved.
(ii) Suppose $(\alpha)$ is not true. Choose $\theta \in \overline{j_E(E)} \backslash j_E(E)$. Then $\theta$ is not $w'$-continuous so $H :=$ Ker $\theta$ is not $w'$-closed. But $\theta$ is $bw'$-continuous by Proposition 3.3. so $H \cap B_{E'}$ is $w'$-closed.

The results of this section yield the existence of polar non-Krein-Šmulian spaces (see §2).

**COROLLARY 3.6.** *If $m$ is a cardinality $\geq \#K$ then $c_0(m)$ is not a Krein-Šmulian space.*

*Proof.* In [2], Exercise 4.N a Banach space $E$ is constructed such that $j_E(E)$ is a proper dense subset of $E''$. From this construction it is easily seen that $\#E = \#\ell^\infty \leq \#K^{\mathbb{N}} = \#K$. Now let $I$ be a set with cardinality $\geq \#K$ and let $\{e_i : i \in I\}$ be the natural orthonormal base of $c_0(I)$. There is a surjection $\{e_i : i \in I\} \to B_E$, it extends to a quotient map $c_0(I) \to E$. Now $E$ is not a Krein-Šmulian space by Theorem 3.1, neither is $c_0(I)$ by Theorem 2.6. (It is not hard to see by looking at the proof of Theorem 2.6 that one can even find a sub*space* $D \subset \ell^\infty(I)$ that is not $w'$-closed while $D \cap B_{\ell^\infty(I)}$ is.)

**COROLLARY 3.7.** *If $m$ is a nonmeasurable cardinality $\geq \#K^K$ then $\ell^\infty(m)$ is not a Krein-Šmulian space.*

*Proof.* In the spirit of the previous proof one constructs a quotient map $\pi : c_0(m) \to \ell^\infty(n)$ where $n = \#K$. By reflexivity ([2], Theorem 4.21) the adjoint $\pi' : c_0(n) \to \ell^\infty(m)$ is an isometry and $\pi'(c_0(n))$ has the WEP in $\ell^\infty(m)$. From [4], Lemma 2.2 we obain that $\pi'(c_0(n))$ is also weakly closed in $\ell^\infty(m)$. By the previous corollary $c_0(n)$ is not a Krein-Šmulian space, neither is $\ell^\infty(m)$ by Theorem 2.7.

**PROBLEM.** Determine the smallest cardinality $m$ for which $c_0(m)$ ($\ell^\infty(m)$ if $\#K$ is nonmeasurable) is not a Krein-Šmulian space.

As a further application we now prove a nonarchimedean version of a classical reflexivity criterion (Theorem 3.8). First we 'dualize' the notion of a polar seminorm as follows. A seminorm $p$ on the dual $E'$ of a locally $K$-convex space $E$ is a *dual seminorm* if there exists an $X \subset E$ such that $p(f) = \sup\{|f(x)| : x \in X\}$ for all $f \in E'$. An easy exercise shows that $p$ is dual if and only if $\{f \in E' : p(f) \leq 1\}$ is $\sigma(E', E)$-closed. Dual seminorms are automatically polar.

**THEOREM 3.8.** *Let $E$ be a pseudoreflexive $K$- Banach space. Then $E$ is reflexive if and only if each polar norm on $E'$ inducing the topology is dual.*

*Proof.* Let $E$ be reflexive and let $\nu$ be a polar norm on $E'$ inducing the topology. Then $\{f \in E' : \nu(f) \leq 1\}$ is weakly closed so, by reflexivity, $w'$-closed. Hence $\nu$ is dual by the above remark. Conversely, suppose each polar norm on $E'$ inducing the topology is dual. To prove reflexivity of $E$ it suffices (by pseudoreflexivity) to show that any $\theta \in E''$ is $w'$-continuous. For each $n \in \mathbb{N}$ the norm $f \mapsto n|\theta(f)| \vee \|f\|$ (where $\| \;\; \|$ is the 'natural' norm on $E'$) is easily seen to be polar and it is obviously equivalent to $\| \;\; \|$. By assumption its closed unit ball

$$B_n := \{f \in E' : |\theta(f)| \leq \frac{1}{n}, \;\; \|f\| \leq 1\}$$

is $w'$-closed. Hence so is $\bigcap_n B_n$ which is Ker $\theta \cap B_{E'}$. By Theorem 3.1 Ker $\theta$ is $w'$-closed implying that $\theta$ is $w'$-continuous.

**Remark.** One also may consider a 'predual form' of the Krein-Šmulian property (compare Definition 1.1, see also Corollary 1.6 (ii)) as follows. A normed space $E$ is PKŠ space if for each absolutely convex edged $A \subset E$:

$$A \cap B \text{ is } w\text{-closed in } B \text{ for each bounded } B \subset E \implies A \text{ is } w\text{-closed.}$$

(Obviously this notion is of no use in classical Banach space theory.) The reader will not have difficulties in proving results about PKŠ spaces similar to the one of KŠ-spaces of this paper. More precisely, we have

(i) A strongly polar normed space is PKŠ.

(ii) Let $E$ be a normed PKŠ space, let $D$ be a closed subspace. Then $E/D$ is PKŠ. If $D$, in addition, is weakly closed and has the WEP then $D$ has PKŠ.

(iii) Let $E$ be a normed space. If $H \subset E$ is a subspace with finite codimension and $H \cap B_E$ is weakly closed then $H$ is weakly closed.

## REFERENCES

[1] N. De Grande-De Kimpe and C. Pérez-García. *Weakly closed subspaces and the Hahn-Banach extension property in p-adic analysis.* Nederl. Akad. Wetensch. Indag. Math. **50** (1988), 253–261.

[2] A.C.M. van Rooij. *Non-Archimedean Functional Analysis.* Marcel Dekker, New York, 1978.

[3] W.H. Schikhof. *The continuous linear image of a p-adic compactoid.* Nederl. Akad. Wetensch. Indag. Math. **51** (1989), 119–123.

[4] W.H. Schikhof. *The complementation property of $\ell^\infty$ in p-adic Banach spaces.* In *p-adic Analysis. Proceedings, Trento 1989*, edited by F. Baldassarri, S. Bosch and B. Dwork, Springer-Verlag (Lect. Notes 1454), Berlin-Heidelberg-New York (1990), 64–79.

[5] J. van Tiel. *Espaces localement K-convexes.* Nederl. Akad. Wetensch. Proc. Ser. A **68** = Indag. Math. **27** (1965), 249–289.

KATHOLIEKE UNIVERSITEIT
MATHEMATISCH INSTITUUT
TOERNOOIVELD
6525 ED NIJMEGEN, THE NETHERLANDS

# TOPOLOGICAL FIELDS AND NONARCHIMEDEAN ANALYSIS

Niel Shell

**ABSTRACT**

The notion of a near normed linear space over a near valued field is introduced. Properties related to the value group of the near norm are considered. The paper begins with a survey of topological fields which is intended as an outline of material useful for studying near normed spaces.

## 1. INTRODUCTION

Nonarchimedean analysis is the study of analogues of theorems in classical analysis that arise when nonarchimedean valued fields (see Definition 5 below) play the role usually played by the real or complex numbers. The earliest branch of nonarchimedean analysis, viz., the theory of nonarchimedean normed linear spaces, illustrates the way a valued field (see Definition 5 and the comments following Definition 5) is substituted for the real or complex field.

**DEFINITION 1.** Let $K$ be a field on which is defined a real valued valuation $|\ \ |$. A nonnegative real valued function $\|\ \ \|$ on a vector space $X$ over $K$ is called a *norm* on $X$ if

(N1) $\|x\| = 0$ if and only if $x = 0$;

(N2) $\|ax\| = |a|\ \|x\|$ for all $a \in K$ and $x \in X$; and

(N3) $\|x+y\| \leq \|x\|+\|y\|$ for all $x, y \in X$.

The norm is called *nonarchimedean* if (3) may be replaced by the stronger condition

(N3′) $\|x+y\| \leq \max(\|x\|, \|y\|)$ for all $x, y \in X$.

If $X$ is a nonzero nonarchimedean normed linear space over $K$, then the inequality $\|(a+b)x\| \leq \max(|a|\ \|x\|, |b|\ \|x\|)$ shows the valuation on $K$ also must be nonarchimedean (i.e., also must satisfy the condition (N3′)).

A natural way to generalize nonarchimedean analysis is to replace a valued field with a more general topological field (see Definition 2 below). For the study of topological vector spaces, for example, a natural definition of topological vector spaces over arbitrary topological fields is obvious. On the other hand, it is not obvious how, if it all, one may define a normed linear space over an arbitrary topological field.

The purpose of this article is to give a survey of the theory of topological fields, describing how special a subclass of the class of topological fields is formed by the valued fields; to introduce a natural generalization of the notion of a normed linear space; and to describe the class of topological fields for which this generalization is suitable. The reader is referred to [15] for complete proofs of the material on topological fields outlined here.

The cardinality of a set $E$ will be denoted by $\#E$. If $A$ and $B$ are subsets of a partially ordered set, we say $A$ is coinitial (cofinal) in $B$ if, for each $b \in B$, there exists $a \in A$ such that $a \leq b$ ($a \geq b$). The subgroup (convex subgroup) generated by a subset $E$ of a group (partially ordered group) will be denoted by $[E]$ ($[E]_{\text{cg}}$). We denote by $\mathbb{Q}$, $\mathbb{R}$, and $\mathbb{C}$ the sets of rational, real and complex numbers, respectively. For any set $E$ we let $E^*$ denote the set $E\backslash\{0\}$. A topology which is neither trivial nor discrete will be called a *proper* topology. Recall that $(G, \leq)$ is called a directed group if it is a partially ordered group such that, for all $x, y \in G$ there exists $z \in G$ such that $z \geq x, y$. For a subset $E$ of a partially ordered additive group $G$ (especially $\mathbb{R}$), $E_{>0}$ ($E_{\geq 0}$) is the set of elements in $E$ that are greater than zero (greater than or equal to zero). The convex subgroup generated by an abelian directed subgroup $H$ of a partially ordered group $G$ equals $\cup_{h\in H}\{g \in G : h^{-1} \leq g \leq h\}$ (see [15, p.189]).

A collection $\mathbf{T}$ of topologies on an arbitrary set $X$ is called *independent* if $\cap U_i \neq \emptyset$ for each finite collection of nonempty sets $\{U_i\}$ such that $U_i \in \mathcal{T}_i$, where $\mathcal{T}_i \in \mathbf{T}$ and $\mathcal{T}_i \neq \mathcal{T}_j$ for $i \neq j$. A collection of topologies induced on a field by nontrivial valuations is independent.

## 2. TOPOLOGICAL FIELDS

DEFINITION 2. We say $\mathcal{T}$ is a ***group topology*** on a group $G$ if $(G, \mathcal{T})$ is a topological group. A group topology on (the additive group of) a ring $R$ is called a ***ring topology*** (and $R$ is called a topological ring) if multiplication is continuous. A ring topology on a field $K$ is called a ***field topology*** (and $K$ is called a topological field) if inversion is a continuous function from $K^*$ to $K^*$.

A group topology on a module $X$ over a topological ring $(R, \mathcal{T})$ is said to be an $(R, \mathcal{T})$-***module topology*** (and $X$ is called an $(R, \mathcal{T})$-topological module) if scalar multiplication is continuous. The terms ***vector space topology*** and ***topological vector space*** are defined analogously.

Suppose $\mathcal{T}$ is a ring topology on a field $K$ and $(X, \mathcal{S})$ is a $(K, \mathcal{T})$-topological vector space. For each fixed $x \in X^*$, let $\mathcal{S}_x = \{U \subset K : Ux \in \mathcal{S}|_{Kx}\}$ and $\mathcal{T}_x = \{Ux : U \in \mathcal{T}\}$. The weak topology induced by

$$\begin{aligned} K &\longrightarrow (Kx, \mathcal{S}|_{Kx}) \\ a &\longmapsto ax \end{aligned}$$

is $\mathcal{S}_x$, and the quotient topology induced by

$$\begin{aligned} (K, \mathcal{T}) &\longrightarrow Kx \\ a &\longmapsto ax \end{aligned}$$

is $\mathcal{T}_x$. Therefore, $\mathcal{S}_x \le \mathcal{T}$ and $\mathcal{S}|_{Kx} \le \mathcal{T}_x$. The topological vector space (and the vector space topology) is called ***straight*** if $\mathcal{S}|_{Kx} = \mathcal{T}_x$ for each $x \in X^*$.

The neighborhood filter at the identity for a group topology $\mathcal{T}$ will be denoted by $\mathcal{B}(\mathcal{T})$.

A group topology determined on a commutative ring $R$ by a neighborhood base $\mathcal{B}$ at zero, is a ring topology if and only if $\mathcal{B}$ satisfies these two conditions:

(B1) Given $U \in \mathcal{B}$, there exists $V \in \mathcal{B}$ such that $VV \subset U$.

(B2) Given $x \in R^*$ and $U \in \mathcal{B}$, there exists $V \in \mathcal{B}$ such that $xV \subset U$.

A ring topology on a field determined by a neighborhood base $\mathcal{B}$ at zero is a field topology if the condition below is satisfied.

(B3) Given $U \in \mathcal{B}$, there exists $V \in \mathcal{B}$ such that $(1+V)^{-1} \subset U$.

For a ring topology on a field with base $\mathcal{B}$, (B3) is equivalent to the following:

(B3′) Given $U \in \mathcal{B}$, there exists $V \in \mathcal{B}$ such that $V/(1+V) \subset U$.

Since the closure of (the singleton) zero and the connected component of zero in a field with a ring topology are ideals, the following important results are immediate:

**THEOREM 1.** (1) *A field with a nontrivial ring topology is Hausdorff.*
(2) *A field with a ring topology is connected or totally disconnected.*

It is easy to verify that the supremum of a collection $\mathbf{T}$ of ring (field) topologies on a field is again a ring (field) topology; and the set of all finite intersections $\cap U_i$, where $U_i \in \mathcal{B}(T_i)$ and $T_i \in \mathbf{T}$ for all $i$, is a base at zero for the supremum. From this it follows that the collections $\mathcal{L}_r(K)$ and $\mathcal{L}_f(K)$ of all ring topologies on $K$ and all field topologies on $K$, respectively, are complete lattices (with containment as inequality).

It might seem natural (and sometimes it is necessary) to consider only field topologies on a field. However, there is sufficient interaction between field and ring topologies to warrent consideration of all ring topologies. One example of this interaction is provided by the theorem below.

**THEOREM 2.** (Gelbaum, Olmstead and Kalish [5], Gould [6], Mutylin [13], Weber [19].) *Let $\mathcal{B}$ be a neighborhood base at zero for a Hausdorff topology on a field $K$. The set*

$$\left\{\frac{B}{1+B\backslash\{-1\}} : B \in \mathcal{B}\right\}$$

*is a neighborhood base at zero for a Hausdorff field topology on $K$. This is the finest field topology on $K$ which is coarser than the given topology.*

Not only is it of interest to consider ring topologies on fields, but it is also of interest to study ring topologies more generally on rings, since the uniform completion of a topological field is, with operations extended by continuity, a topological ring, but not necessarily a topological field. For example, the completion of a field with respect to the supremum of an independent collection of ring topologies is topologically isomorphic to the direct product of the completions of the field with respect to the topologies in the collection; thus the completion with respect to the supremum has divisors of zero (provided the collection of topologies has more than one element). It is also possible for the completion to be an integral domain which is not a field (see [21]).

Among topological fields we will see (Theorem 9 and the comments following Theorem 9) that the rank one nonarchimedean valued fields are the minimal locally bounded nonarchimedean fields with a nonzero topological nilpotent. All the properties of topological fields involved in this statement are discussed below.

By a *minimal* field we mean a field that has a ring topology which is minimal among all nontrivial ring topologies on the field. It is not difficult to prove that a topology induced by a valuation is minimal. No other examples of minimal fields are known (although Hartmann [7] has given an example of a minimal skew field topology that is not induced by a valuation). According to Theorem 2, a minimal ring topology is a field topology. It may be proved that the completion of a minimal field is again a topological field.

An element $x$ in a topological ring is *topologically nilpotent* if $x^n$ converges to zero.

A uniform space is called *nonarchimedean* if it has a base of entourages $U$ such that $U \circ U \subset U$. In particular, an abelian topological group with its natural uniformity is nonarchimedean if and only if it has a base at the identity consisting of subgroups. We will say a topological ring is nonarchimedean if its additive group is nonarchimedean.

## 3. LOCALLY BOUNDED FIELDS

A subset $B$ of a topological module $X$ over a commutative topological ring $R$ is called *bounded* if, for each neighborhood $U$ of zero in $X$, there exists a neighborhood $V$ of zero in $R$ such that $VB \subset U$. A subset of a commutative topological ring is called bounded if it is bounded when the ring is viewed as a module over itself (with the ring multiplication as scalar multiplication). In a field $K$ with a proper ring topology, a set $B$ is bounded if and only if, for each neighborhood $U$ of zero, there exists an element $x \in K^*$ such that $xB \subset U$. One readily sees that a subset $B$ of a valued field $(K, v)$ is bounded if and only if there exists an element $g$ of the value group $v(K^*)$ such that $v(b) \leq g$ for all $b \in B$. A ring topology (or a topological ring) is *locally bounded* if it has a bounded neighborhood of zero. Obviously a valued field is locally bounded.

The topology of a locally bounded field can be described in an especially easy way. If, for a subset $U$ of a field $K$, the collection $\{xU : x \in K^*\}$ forms a neighborhood base at zero for a ring topology we say that $U$ induces this ring topology and denote it by $\mathcal{T}_U$.

**THEOREM 3.** *Each topology $\mathcal{T}_U$ induced on a field $K$ by a set $U$ is locally bounded, and in this topology $U$ is a bounded neighborhood of zero. Conversely, for each bounded neighborhood $U$ of zero in a locally bounded topology $\mathcal{T}$, ($U$ induces a ring topology $\mathcal{T}_U$ and) $\mathcal{T}_U = \mathcal{T}$.*

This convenient description for the topology can be improved. The neighborhood $U$ used to induce the topology can be chosen to be "nearly a valuation ring", in a sense made precise by the following definition.

**DEFINITION 3.** Let $K$ be a field. A subset $A$ of $K$ is called a *preorder* if the following conditions are satisfied:

(O1) $0, 1 \in A$.

(O2) $AA \subset A$.

(O3) $K^* = A^*(A^*)^{-1}$.

(O4) There exists an element $z \in A^*$ such that $z(A+A) \subset A$.

An element $z$ for which (O4) is satisfied is called an *addiator*.

An element $d \in K^*$ will be called a *dominator* of a subset $B$ of $K$ if $B+B \subset dB$.

A preorder $A$ is called a *near order* if

(O1′) $0, -1 \in A$.

If a preorder $A$ also satisfies

(O5) for each $x \in K^*$, $x \in A$ or $x^{-1} \in A$,

then ($A$ is a near order and) $A$ is called a *valuation* near order (*Vollfastordnung* in Kowalsky and Dürbaum [12]).

A near order $A$ is an order of $K$ (i.e. $A$ is a subring of $K$ with identity and with quotient $K$) if 1 is an addiator. A valuation order of $K$ is a valuation ring of $K$.

Different authors use slightly different axioms in the definition of near order and slightly different translations of the original term *Fastordnung* introduced by Kowalsky and Dürbaum [12].

For each bounded neighborhood $U$ of zero in a locally bounded topology $\mathcal{T}$ on a field $K$, the set $A = \{x \in K : xU \subset U\}$ is a preorder inducing $\mathcal{T}$ (and, if $U$ is symmetric, $A$ is a near order). Clearly, the valuation ring of a valuation is a valuation near order inducing the topology of the valuation. The relationship between locally bounded topologies, preorders, and bounded neighborhoods is illustrated with two simple examples.

**EXAMPLE 1.** The interval$(-1,1/2] \cup \{1\}$, which we will denote by $U$, is a bounded neighborhood in $\mathbb{R}$ with the usual topology;

Clearly, $U$ induces the usual topology, i.e., $\{xU : x \in \mathbb{R}^*\}$ is a base at zero for the usual topology. Define $A := \{x \in \mathbb{R} : xU \subset U\}$. Then $A = [-1/2, 1/2] \cup \{1\}$, and $A$ is a preorder; $\{xA : x \in \mathbb{R}^*\}$ is also a neighborhood base at zero for the usual topology. Even though the usual topology is induced by a valuation (the usual absolute value), the preorder $A$ is not a valuation near order, and in fact not a near order at all. However, $[-1,1]$ is a valuation near order inducing the usual topology.

One readily checks that $A+A = [-1,3/2] \cup \{2\}$. If $d > 0$ is a dominator of $A$, then $[-1/2, 3/2] \subset dA = [-d/2, d/2] \cup \{d\}$ and $d \geq 3$. Since, also, $2 \in dA$, we have $d \geq 4$. Thus the numbers with absolute value greater than or equal to 4 are the dominators of $A$, or, equivalently, the numbers with absolute value less than or equal to $1/4$ are the addiators of $A$. One easily sees that the numbers with absolute value greater than or equal to 2 are the dominators of $[-1,1]$.

**EXAMPLE 2.** The ring $\mathbb{Z}$ of integers is an order of $\mathbb{Q}$ and, thus, induces a ring topology on $\mathbb{Q}$. Since $a\mathbb{Z} \subset (a/b)\mathbb{Z}$, whenever $a$ and $b$ are nonzero integers, the nonzero ideals of $\mathbb{Z}$ form a base at zero for this ring topology. If $a/b$ is an irreducible representation of a dominator of $\mathbb{Z}$, then, since $1 \in \mathbb{Z}+\mathbb{Z} \subset a\mathbb{Z}/b$, we see $a = \pm 1$. Thus, the reciprocals of nonzero integers are the dominators of $\mathbb{Z}$.

$\mathcal{T}_{\mathbb{Z}}$ is not a field topology. In fact, we observe that, in a sense, inversion is as discontinuous as possible. If $\{x_n\}$ is a sequence such that $x_n$ converges to 1 and $x_n^{-1}$ converges to 1, then $x_n - 1$ is eventually in the neighborhood $\mathbb{Z}$ of zero, or, equivalently, all but finitely many terms of the sequence ($x_n - 1$ and hence) $x_n$ are integers. Similarly, all but finitely many of the terms of $x_n^{-1}$ are integers. These observations together imply $x_n = \pm 1$ for all but finitely many $n$. Certainly then, $x_n = 1$ for all but finitely many $n$. An easy consequence of this result is that, if $x_n$ converges to $x$ and $x_n^{-1}$ converges to $x^{-1}$, then $x_n$ is eventually equal to $x$: apply the above result to the sequence $x^{-1}x_n$.

From the comments following Theorem 1 it follows that $\mathbb{Z}$ is not a neighborhood of zero in the supremum of all $p$-adic topologies on $\mathbb{Q}$. However, each neighborhood of zero in the supremum of all $p$-adic topologies contains an ideal of $\mathbb{Z}$. Therefore, $\mathcal{T}_{\mathbb{Z}}$ is strictly finer than the supremum of all $p$-adic topologies. It is not difficult to show that the supremum of all $p$-adic topologies is not locally bounded (see [11] or [15, Theorem 4.5.5]).

We describe the completion of $(\mathbb{Q}, \mathcal{T}_{\mathbb{Z}})$. Let $\mathbb{Q}_p$ denote the $p$-adic field, i.e., the completion of $\mathbb{Q}$ with the $p$-adic valuation $| \quad |_p$, and let $\mathbb{P}$ denote the set of all prime natural numbers. Let $\widehat{\mathbb{Q}}$ denote the subset of all elements $\{x_p\}$ in $\prod_{p\in\mathbb{P}} \mathbb{Q}_p$ such that $|x_p|_p \leq 1$ for all but finitely many $p$. Let $B$ be the compact topological ring $\prod_{p\in\mathbb{P}}\{x_p \in \mathbb{Q}_p : |x_p|_p \leq 1\}$ with the product topology (with each factor being viewed as subspace of $\mathbb{Q}_p$). The collection of all neighborhoods of zero in $B$ is a neighborhood base for a complete ring topology, which we call $\widehat{\mathcal{T}}_{\mathbb{Z}}$, on $\widehat{\mathbb{Q}}$. The map from $(\mathbb{Q}, \mathcal{T}_{\mathbb{Z}})$ to $(\widehat{\mathbb{Q}}, \widehat{\mathcal{T}}_{\mathbb{Z}})$ which assigns to each $x$ the element of the product with components all equal to $x$ is a (well-defined) topological isomorphism onto a dense subfield in $\widehat{\mathbb{Q}}$. Thus $(\widehat{\mathbb{Q}}, \widehat{\mathcal{T}}_{\mathbb{Z}})$ is the completion of $(\mathbb{Q}, \mathcal{T}_{\mathbb{Z}})$. Similarly, $\prod_{p\in\mathbb{P}} \widehat{\mathbb{Q}}_p$, the entire product space with the product topology, is the completion of $\mathbb{Q}$ with the supremum of all $p$-adic topologies.

The following theorem tells exactly which locally bounded fields are nonarchimedean:

**THEOREM 4.** ([20]) *A ring topology on a field is induced by an order if and only if it is locally bounded and nonarchimedean.*

**DEFINITION 4.** A nonnegative real valued function $\| \quad \|$ on a ring $R$ is called a *ring seminorm* if, for all $x, y \in R$,

(N1′) $\|0\| = 0$;

(N2′) $\|-x\| = \|x\|$;

(N3) $\|x+y\| \leq \|x\|+\|y\|$

(N4) $\|xy\| \leq \|x\| \, \|y\|$.

A ring seminorm is called a *ring norm* if only the zero element has norm zero.

Since the elements with seminorm zero form an ideal, a nonzero ring seminorm on a field is a ring norm. The following result is easy to show: If $F$ is a field with a ring norm which, when restricted to a subfied $K$, is an absolute value, then $F$ is a normed linear space over $K$. A ring norm on a field induces a topology in an obvious way (the spheres form a base); a topology induced by a norm is a field topology. The topologies on a field induced by ring norms can be characterized:

**THEOREM 5.** ([2]) *Let $\mathcal{T}$ be a proper ring topology on a field $K$. The following statements are equivalent:*

(1) *$\mathcal{T}$ is induced by a ring norm.*

(2) *$\mathcal{T}$ is locally bounded and there exists a nonzero topologically nilpotent element.*

(3) *$\mathcal{T}$ is locally bounded and the set of topologically nilpotent elements form a neighborhood of zero.*

For distrinct primes $p$ and $q$, the supremum of the $p$-adic topology and the $q$-adic topology on $\mathbb{Q}$ is a topology which is induced by a norm but not by a valuation. According to our earlier comments the completion of this normed ring has divisors of zero. Mutylin [13] has given an example of a *complete* field topology which is induced by a ring norm but not by a valuation.

The following easily proved result indicates the class of fields on which there exists a proper locally bounded topology. In the theorem below we refer to an algebraic extension field of a finite field as *absolutely algebraic.*

**THEOREM 6.** *Let $K$ be a field. The following statements are equivalent.*

(1) *$K$ is absolutely algebraic.*

(2) *The only rank one valuation on $K$ is the trivial valuation.*

(3) *The only valuation on $K$ is the trivial valuation.*

(4) *There are no proper locally bounded topologies on $K$.*

An obvious consequence of Theorem 6 is that there is a proper ring topology on each uncountable field. A considerably deeper result is that there are proper ring topologies on each infinite field (see [9] and [1]). In fact, there are $2^{(2^{\#K})}$ distinct ring topologies on each infinite field $K$ (see [10]). Theorem 1(1) implies that there are no proper ring topologies on a finite field.

The natural correspondence between valuations rings and equivalence classes of valuations has a natural extension to preorders.

**DEFINITION 5.** Let $G$ be a directed abelian multiplicative group to which we adjoin a least element 0 (with multiplication extended to $G \cup \{0\}$ by defining $0 \cdot g = g \cdot 0 = 0$ for every $g \in G \cup \{0\}$).

A surjective function $v : K \to G \cup \{0\}$ is called a ***directed homomorphism*** on the field $K$ if the following conditions are satisfied:

(V1) $v(x) = 0$ if and only if $x = 0$.

(V2) $v(xy) = v(x)v(y)$ for all $x, y \in K$.

A directed homomorphism is called a ***prevaluation*** if also

(V3) there exists $\lambda \in G$ such that $v(x)$, $v(y) \leq g$ implies $v(x+y) \leq \lambda g$.

A prevaluation is called a ***near valuation*** if

(V4) $v(-1) = 1$.

An element $\lambda$ for which (V3) is satisfied is called a ***dominator*** of $v$. If $G$ is totally ordered, $v$ is called a ***valuation.*** A prevaluation $v$ is called ***nonarchimedean*** if the identity of $G$ is a dominator. The group $G$ is called the ***value group*** of the prevaluation. A prevaluation is called ***discrete*** if its value group is infinite cyclic. A valuation $v$ is called a ***real*** valuation if its value group is an ordered subgroup of $\mathbb{R}_{>0}$.

A ***prevalued field*** is a pair $(K, v)$, where $K$ is a field and $v$ is a prevaluation on $K$.

A function $| \ | : K \to \mathbb{R}_{\geq 0}$ on a field $K$ is called an ***absolute value*** if, for all $x, y \in K$,

(AV1) $|x| = 0$ if and only if $x = 0$;

(AV2) $|xy| = |x|\,|y|$;

(AV3) $|x+y| \leq |x|+|y|$.

Each dominator is greater than or equal to one: $1 = v(1+0) \leq \lambda$. Each absolute value is a valuation.

The definition of valuation used here (due to Kowalsky and Dürbaum [12]) is slightly more general than the usual definition. For example, we may define a rank two valuation on $\mathbb{Q}(z)$ which induces the usual topology: Choose $z \in \mathbb{C}\backslash\mathbb{R}$ such that the usual absolute value of $z$ is one and $z$ is transcendental over $\mathbb{Q}$. Let $v(x) = (|x|_\infty, |x|_z)$, where $| \ |_\infty$ is the usual absolute value and $| \ |_z$ is the $z$-adic valuation and $|\mathbb{Q}(z)^*|_\infty \times |\mathbb{Q}(z)^*|_z$ is ordered lexicographically. The terms absolute value and nonarchimedean valuation as defined here have their usual meanings.

For a subset $E$ of a ring, we let $E^\times = \{x \in E : x^{-1} \in E\}$. For each preorder $A$ of a field $K$, we define a partial order on $K^*/A^\times$ by defining $xA \leq yA$ if $xy^{-1} \in A$. The canonical homomorphism $v$ onto this quotient (with $v(0) = 0$) is a prevaluation; $\{x : v(x) \leq 1\} = A$; and $d$ is a dominator of $A$ if and only if $v(d)$ is a dominator of $v$.

We illustrate the relation between prevaluations and locally bounded topologies by continuing the discussion of Examples 1 and 2 in.

**EXAMPLE 1** (continued). Since $A^{\times} = \{1\}$, the induced prevaluation $v_A$ is the identity map with $\mathbb{R}^*$ considered as a directed multiplicative group with the order

$$a <_A b \iff |a| \leq \tfrac{1}{2}|b|.$$

The value group of $v_A$ is not totally ordered (e.g., 2 and 3 are not comparable), so $v_A$ is not a valuation. Since $v_A(-1) = -1$, which is not equal to (in fact, not comparable to) 1, $v_A$ is not a near valuation. The equality $\{x : v_A(x) \leq 1\} = A$ illustrates the extension to prevaluations and preorders of the usual correspondence between valuations and valuation rings. According to the correspondence indicated above between dominators of $A$ and dominators of $v_A$, the elements with absolute value greater than or equal to 4 are the dominators of $v_A$.

The value group of $v_{[-1,1]}$ consists of all pairs $\{\pm x\}$, $x \neq 0$. The map $\{\pm x\} \mapsto |x|$ is an order preserving isomorphism (with order preserving inverse) of the value group of $v_{[-1,1]}$ onto $\mathbb{R}_{>0}$ with the usual order. If we use this isomorphism to identify the value group with $\mathbb{R}_{>0}$, then $v_{[-1,1]}$ is the usual absolute value.

**EXAMPLE 2** (continued). As above, the mapping $\{\pm a\} \mapsto |a|$ may be used to identify the value group of $v_{\mathbb{Z}}$ with $\mathbb{Q}_{>0}$ ordered by the relation

$$a \leq_{\mathbb{Z}} b \iff a/b \in \mathbb{Z}.$$

With this identification, $v_{\mathbb{Z}}(x) = |x|$ (and we remember to consider $\mathbb{R}_{>0}$ ordered as above). The reciprocals of positive integers are the dominators of $v_{\mathbb{Z}}$. Thus, this near valuation is nonarchimedean.

## 4. TYPE *V* FIELDS

**DEFINITION 6.** A subset of a field $(K, \mathcal{T})$ with a ring topology is said to be *bounded away from zero* if it is disjoint from some neighborhood of zero; $K$ (also $\mathcal{T}$) is said to be of *type V* (or *locally retrobounded*) if the inverse of each set bounded away from zero is bounded.

Adjoining a point at infinity helps us interpret our results geometrically. Let $\mathcal{T}$ be a ring topology on a field $K$, and let $K_\infty = K \cup \{\infty\}$, where $\infty$ is any element not in $K$. We let

$$\mathcal{T}(\infty) = \mathcal{T} \cup \{K_\infty \backslash B : B \text{ is } \mathcal{T}\text{-closed and bounded}\}.$$

We extend the algebraic operations by defining $0^{-1} = \infty$, $\infty^{-1} = 0$, and $a \cdot \infty = \infty \cdot a = \infty$ for $a \in (K_\infty)^*$.

**THEOREM 7.** (Shilkret [16], Shell [14].) *Let $(K, \mathcal{T})$ be a field with a proper ring topology.*

(1) *$\mathcal{T}(\infty)$ is a topology inducing $\mathcal{T}$ on $K$; a subset of $K_\infty$ is a neighborhood of $\infty$ if and only if its complement is a $\mathcal{T}$-bounded subset of $K$.*
(2) *$(K_\infty, \mathcal{T}(\infty))$ is Hausdorff if and only if $(K, \mathcal{T})$ is locally bounded.*
(3) *$(K_\infty, \mathcal{T}(\infty))$ is a compact Hausdorff space if and only if $(K, \mathcal{T})$ is a locally compact Hausdorff space. In this case (Heine-Borel) $(K_\infty, \mathcal{T}(\infty))$ is the one point compactification of $K$ (i.e., a set is compact if and only if it is closed and bounded).*

**THEOREM 8.** (Kaplansky [8], Shell [14]) *Let $\mathcal{T}$ be a proper ring topology on a field $K$. The following statements are equivalent:*
(1) *$\mathcal{T}$ is of type $V$.*
(2) *Inversion is continuous at infinity.*
(3) *$(K, \mathcal{T})$ is locally bounded and multiplication on $[(K_\infty)^*]^2$ is continuous at $(\infty, \infty)$.*

The concepts now introduced allow us to characterize the ring topologies on a field that are induced by a valuation:

**THEOREM 9.** (Turyn [17], Fleischer [3], Fleischer [4], Kowalsky and Dürbaum [12], Kowalsky [11]) *Let $\mathcal{T}$ be a Hausdorff ring topology on the field $K$. The following statements are equivalent.*
(1) *$\mathcal{T}$ is induced by a valuation.*
(2) *$(K, \mathcal{T})$ is topologically isomorphic to a subfield of $(\mathbb{C}, \mathcal{T}_\infty)$ or $\mathcal{T}$ is induced by a nonarchimedean valuation.*
(3) *$\mathcal{T}$ is of type $V$.*
(4) *$\mathcal{T}$ is locally bounded and minimal.*

A topology induced by a nontrivial valuation can be induced by an absolute value if and only if there exists a nonzero topologically nilpotent element. All locally compact ring topologies on fields are induced by absolute values (see, e.g., [18, Section 27]).

Theorem 9 is remarkable not only for its importance in the theory of topological fields, but also because it says that some very different appearing conditions are equivalent: (1) and (2) state that the topology is induced by "very nice functions"; (3) is a statement about continuity of multiplication and inversion at infinity; (4) says the topology is "fairly nice" and is "in a nice position" in the lattice of ring topologies.

## 5. NEAR NORMED LINEAR SPACES

DEFINITION 7. Let $(K, v)$ be a near valued field, and let $G$ be a partially ordered group containing $v(K^*)$ as an ordered subgroup. A function $N$ from a $K$-vector space $X$ into $G$ is called a *near norm* on $X$ if it satisfies the following conditions:

(NN0) $v(K^*)$ is coinitial and cofinal in $N(X^*)$.

(NN1) $N(x) = 0$ if and only if $x = 0$.

(NN2) $N(ax) = v(a)N(x)$ for all $a \in K$ and $x \in X$.

(NN3) There exists $\mu \in [v(K^*)]_{cg}$ such that $N(x), N(y) \leq g$ and $g \in [v(K^*)]_{cg}$ implies $N(x+y) \leq \mu g$.

An element $\mu$ for which (NN3) is satisfied is called a *dominator* of $N$. The near norm $N$ is called *nonarchimedean* if the identity is a dominator.

In [18], a near norm $N$ on a nonarchimedean valued field $(K, v)$ (with $N$ viewed as having $[v(K^*)]_{cg}$ as codomain) is called a *standard*.

By replacing both a dominator of $v$ and a dominator of $N$ by an element of $v(K^*)$ bigger than both dominators, we may choose a value which is a dominator for both $v$ and $N$.

LEMMA 10.1. *For $A \subset G$, where $G$ is a partially ordered group having a value group $v(K^*)$ as an ordered subgroup, the following statements are equivalent:*

(1) *$v(K^*)$ is coinitial and cofinal in $A$.*

(2) *$v(K^*)$ is coinitial in $[A]$.*

(3) $v(K^*)$ *is coinitial in* $A \cup A^{-1}$.

(4) $A \subset [v(K^*)]_{cg}$.

THEOREM 10. *The group* $[NX^*)]$, *where* $N$ *is a near norm on* $X$, *is directed*

*Proof.* A partially ordered group with a coinitial directed subgroup is easily seen to be directed.

LEMMA 11.1. *Assume only that* $N$ *satisfies* (NN2). $N(X^*)$ *is coinitial (cofinal) in* $v(K^*)$ *if and only if there exists* $x \in X^*$ *such that* $N(x) \leq 1$ ($N(x) \geq 1$).

THEOREM 11. *For each near norm* $N$ *on a vector space* $X$ *over* $(K, v)$, $N(X^*)$ *is coinitial and cofinal in* $v(K^*)$.

*Proof.* For $x \in X^*$, choose $a \in K^*$ such that $v(a) \leq N(x)$. Then $1 = v(a^{-1}a) \leq N(a^{-1}x)$, so $N(X^*)$ is cofinal in $v(K^*)$ by Lemma 11.1.

COROLLARY 11.1. *For each near normed space* $(X, N)$ *over* $(K, v)$,

$$[v(K^*)]_{cg} = [N(X^*)]_{cg} = \bigcup_{a \in v(K^*)} \{g \in G : v(a^{-1}) \leq g \leq v(a)\};$$

*and* $[v(K^*)]_{cg}$ *is directed.*

*Proof.* The displayed inequalities follow immediately from Theorems 10 and 11 and the description of generated convex subgroups given in the introduction. Clearly the convex subgroup generated by a directed subgroup is directed.

The group $[N(X^*)]_{cg}$ is called the ***value group*** of $N$ in $G$. As Example 3 below shows, the value group of $N$ depends not only on $v(K^*)$ and $N(X^*)$, but also on the group $G$ in which these sets are embedded.

For each function $N$ which is a norm, a near valuation or a near norm, we let

$$S_g(x) = \{y : N(y-x) < g\} \qquad C_p(x) = \{y : N(y-x) \leq g\},$$

and call these sets *spheres*. By $S_g(C_g)$ we mean $S_g(0)$ $(C_g(0))$.

EXAMPLE 3. Let $(K, v)$ be a rank one valued field whose value group is a proper dense subgroup of $\mathbb{R}_{>0}$, and let $X = K$ and $N = v$. The value group of $N$ is $v(K^*)$ if $G = v(K^*)$ and the value group of $N$ is $\mathbb{R}_{>0}$ if $G = \mathbb{R}_{>0}$. For $r \in (\mathbb{R}_{>0}) \backslash v(K^*)$, the set $\{x \in X : N(x) < r\}$ is a sphere when $N$ is viewed as a mapping into $\mathbb{R}_{>0}$, but is not a sphere when $N$ is viewed as a mapping into $v(K^*)$.

One readily verifies that, if $N$ is a near norm on $X$, then the collection $\{S_g : g \in N(X^*)\}$ is a neighborhood base at zero for a $(K, v)$-vector space topology on $X$. The sets $\{S_g : g \in [N(X^*)]_{\text{cg}}\}$ form a base at zero for the same topology. We call this the topology induced by $N$ and assume that a near normed space is topologized with the induced topology unless stated otherwise.

If $N$ is a near norm with dominator $\mu$ and $N(y-x) \leq N(x)$, then $N(y) \leq \mu N(x)$ (cf. [15, p.14]).

It is clear that the spheres in a nonarchimedean near normed space are additive groups. From the inequality immediately above, it follows that these spheres are open. Thus, the additive group of a nonarchimedean near normed space is a nonarchimedean topological group. In particular, nonarchimedean near valued fields (which may be considered as one dimensional near normed spaces over themselves) are nonarchimedean (when considered as additive topological groups).

**EXAMPLE 4.** Let $|\ \ |$ be a nonarchimedean absolute value on a field $K$; let $\mathbb{R}(t)$ be an ordered field with $t$ larger than every real number; and let $X = K^2$. The function

$$\begin{aligned} N : X &\longrightarrow \mathbb{R}(t)_{\geq 0} \\ (a, b) &\longmapsto \max(|a|t, |b|) \end{aligned}$$

satisfies (NN1)-(NN3), but not (NN0). The spheres about zero in $X$ form a neighborhood base at zero for a nonarchimedean group topology with respect to which $\{0\} \times K$ is an open subgroup. However, according to our comments at the beginning of Section 2, $X$ is not a topological vector space over $(K, |\ \ |)$ because the subspace $Y$ generated by $(1, 0)$ is discrete.

The restriction of $N$ to $Y$ also satisfies (NN1)-(NN3), but not (NN0). However, the spheres $S_g$, $g \in N(Y^*)$, induce a topology that *does* make $Y$ a $(K, |\ \ |)$-topological vector space.

**EXAMPLE 5.** For each near ordered field $(K, v)$ whose value group is lattice ordered, the function $N(a_1, \ldots, a_n) = \sup v(a_i)$ is a norm on $K^n$ which induces the product topology.

It is natural to ask which topological vector spaces over a near valued field $(K, v)$ are induced by a near norm. If $v(K^*)$ is not lattice ordered, it is not even obvious that $K^n$ is near normable for $n > 1$.

A map $\sqcup : (G \cup \{0\}) \times (G \cup \{0\}) \longrightarrow G \cup \{0\}$, where $G$ is a directed group (and a least element 0 is adjointed to $G$), is called a *near maximum* if the following conditions are satisfied for all $g, h, k \in G \cup \{0\}$:

(Q1) $g \sqcup h = h \sqcup g$.

(Q2) $(g \sqcup h) \sqcup k) = g \sqcup (h \sqcup k)$.

(Q3) $(kg) \sqcup (kh) = k(g \sqcup h)$.

(Q4) $g, h \leq g \sqcup h$.

(Q5) There exists $\mu \in G$ such that $g \sqcup h \leq \mu g$ or $g \sqcup h \leq \mu h$ for all $g, h \in G$.

An element $\mu$ for which (Q5) is satisfied will be called a dominator of $\sqcup$.

A dominator of a near maximum is greater than or equal to 1: $1 \leq 1 \sqcup 1 \leq \mu$; 1 is a dominator of a near maximum $\sqcup$ on $G$ if and only if $G$ is totally ordered and $g \sqcup h = \max(g, h)$ for all $g, h \in G$. Since $0 \cdot g = 0$, (Q3) implies $0 \sqcup 0 = 0$. By induction, $\sqcup_{i=1}^{n+1} g_i \leq \mu^n g_j$ for some $j$.

If $\sqcup$ is a near maximum on the group $v(K^*)$, where $v$ is a near valuation, then $N(a_1, \ldots, a_n) = \sqcup_i v(a_i)$ is a near norm inducing the product topology on $K^n$.

**EXAMPLE 6.** Let $K_n = \mathbb{Q}(6^{(1/2^n)})$, and let $K = \cup_{n=0}^{\infty} K_n$. Let $|\ \ |_2$ and $|\ \ |_3$ denote the unique extensions of the 2-adic and 3-adic valuations, respectively, to $K$. Then $v(a) = (|a|_2, |a|_3)$ on the product of the value groups with the order $(g_1, h_1) < (g_2, h_2)$ if $g_1 < g_2$ and $h_1 < h_2$ is a near valuation inducing the supremum of the 2-adic and 3-adic topologies on $K$. The value group is not lattice ordered. However,

$$(g_1, h_1) \sqcup (g_2, h_2) = \lambda(\max(g_1, g_2), \max(h_1, h_2))$$

(where $(0, 0)$ is taken as the least element adjointed to $v(K^*)^2$) is a near maximum for any fixed element $\lambda > 1$.

## REFERENCES

[1] V.I. Arnautov. *On topologization of countable rings.* Sibirsk Mat. Zh. (Siberian Math. J.) **9** (1968), 1251–1261 (939–946). MR **38** #4528.

[2] P.M. Cohn. *An invariant characterization of pseudo-valuations on a field.* Proc. Cambridge Philos. Soc. **50** (1954), 159–177. MR **16** p.214.

[3] I. Fleischer. *Sur les corps topologiques et les valuations.* C.R. Acad. Sci. Paris Sér. A-B **236** (1953), 1320–1322. MR **14** p.720.

[4] I. Fleischer. *Sur les corps localement bornés.* C.R. Acad. Sci. Paris Sér. A-B **237** (1953), 546–548. MR **15** p.195.

[5] B. Gelbaum, G.K. Kalish and J.H. Olmstead. *On the embedding of topological semigroups and integral domains.* Proc. Amer. Math. Soc. **2** (1951), 807–821. MR **13** p.206.

[6] G.G. Gould. *Locally unbounded topological fields and box topologies on products of vector spaces.* J. London Math. Soc. **36** (1961), 273–281. MR **24** #A413.

[7] P. Hartmann. *Stellen und Topologien von Schiefkörpen und Alternativkörpern.* Arch. Math. (Basel) **51** (1988), 274–282. MR **89**k:12013.

[8] I. Kaplansky. *Polynomials in topological fields.* Bull. Amer. Math. Soc. **54** (1948), 909–916. MR **10** p.280.

[9] J.O. Kiltinen. *Inductive ring topologies.* Trans. Amer. Math. Soc. **134** (1968), 149–169. MR **37** #4054.

[10] J.O. Kiltinen. *On the number of field topologies on an infinite field.* Proc. Amer. Math. Soc. 40 (1973), 30–36. MR **47** #6667.

[11] H.-J. Kowalsky. *Beiträge zur topologischen algebra.* Math. Nachr. **11** (1954), 143–185. MR **15** p.774.

[12] H.J. Kowalsky and H. Dürbaum. *Arithmetische Kennzeichnung von Korpertopologien.* J. Reine Angew. Math. **191** (1953), 135–152. MR **15** p.98.

[13] A.F. Mutylin. *Completely simple commutative topological rings.* Mat. Zametki (Math. Notes.) **5**(5) (1969), 161–171 (100–105). MR **39** #4232.

[14] N. Shell. *Point at infinity.* Bull. Austral. Math. Soc. **38** (1988), 321–323.

[15] N. Shell. *Topological Fields and Near Valuations.* Marcel Dekker, New York, 1990.

[16] N. Shilkret (Shell). *One point adjunctions.* Math. Student **42** (1974), 99. MR **53** #1508.

[17] R. Turyn. Undergraduate thesis, Harvard, 1951.

[18] S. Warner. *Topological Fields.* North-Holland, Amsterdam 1989.

[19] H. Weber. *Ringtopologien auf Z and Q.* Math. Z. **155** (1977), 287–298. MR **56** #5517.

[20] D. Zelinsky. *Topological characterization of fields with valuations.* Bull. Amer. Math. Soc. **54** (1948), 1145–1150. MR **10** p.426.

[21] R. Zobel. *Direkte Gruppen- und Ringtopologien.* Dissertation, Tech.Univ. Braunschweig, 1973.

THE CITY COLLEGE OF NEW YORK (CUNY)
CONVENT AVENUE AT 138TH STREET
NEW YORK, NY 10031, USA

# OPEN PROBLEMS

A.C.M. van Rooij and W.H. Schikhof

It was during the Conference when the idea came up to call upon our colleagues to attack open problems that so far had gotten little attention. The result is a selection of 19 problems all centered around Banach space theory. We do not claim that they reflect the state of the art in $p$-adic Functional Analysis. In fact, in many papers, including the ones of these Proceedings, one may find questions that are at least equally interesting. We also wish to point out that —as far as we know— none of the problems stated in [5] has been solved yet!

## NOTATIONS & TERMINOLOGY

We follow the conventions of [5]. Throughout $K$ is a nonarchimedean nontrivially valued field that is complete under the metric induced by the valuation $|\ \ |$. Norms and seminorms are non-archimedean. $E, F$ are $K$-Banach spaces; $E \oplus F$ is the orthogonal direct sum of $E$ and $F$; the formula $E \sim F$ means that $E, F$ are isometrically isomorphic. The 'open' ('closed') ball with radius $r$ and center $a$ is denoted $B(a, r^-)$ ($B(a, r)$). For $a \in E$ we write $[a]$ for the linear subspace generated by $a$. An absolutely convex subset $A$ of $E$ is *edged* if for each $a \in E$ the set $\{|\lambda| : \lambda \in K,\ \lambda a \in A\}$ is closed in $|K|$.

## 1. FINITE DIMENSIONAL PRIME DECOMPOSITION

Let us call an $E \neq \{0\}$ *prime* if there do not exist non-zero Banach spaces $E_1, E_2$ such that $E \sim E_1 \oplus E_2$. For a finite dimensional Banach space $E$, $E \neq \{0\}$, clearly there exist prime spaces $E_1, \ldots E_N$ such that $E \sim E_1 \oplus \ldots \oplus E_N$.

**Problem.** *Is the prime decomposition of a finite dimensional Banach space in the natural sense unique?*

(One can show: if $E, F$ are finite dimensional and $K \oplus E \sim K \oplus F$ then $E \sim F$. Also, if $E_1, \ldots E_N$ are prime, $D$ is a Banach space and $E_1 \oplus \ldots \oplus E_N \sim K \oplus D$ then $E_n \sim K$ for some $n$.)

**Problem.** *If $E$ is prime, must $\{\|x\| : x \in E,\ x \neq 0\}$ be just one coset of the value group of $K$ in $(0, \infty)$?*

## 2. NORMS ON $K^2$

Let $K$ not be spherically complete. Let $\mathcal{N}$ be the set of all norms $\nu$ on $K^2$ for which

(1) $\nu(x) \in |K| \qquad (x \in K^2)$;

(2) $K^2$ has no base that is orthogonal relative to $\nu$.

For $\mu, \nu \in \mathcal{N}$ write

$$\mu \sim \nu$$

if there is a linear bijection $A : K^2 \to K^2$ such that $\mu = \nu \circ A$.

**Problem.** *Describe the set of all $\sim$-equivalence classes. E.g., can it be finite?*

The following consideration may be of use.

By a *hole* we mean a maximal chain of balls of $K$ with empty intersection. From a hole we can make other holes by translation and by multiplication with a nonzero scalar. Let us call two holes, $\Omega$ and $\Omega'$ *equivalent,*

$$\Omega \sim \Omega'$$

if there exist $\alpha \in K$ and $\beta \in K\backslash\{0\}$ for which $\Omega' = \alpha + \beta\Omega$.

A hole $\Omega$ determines a norm $\nu_\Omega$ on $K^2$ in the following fashion. Let $|\Omega|$ be the infimum of the radii of the balls belonging to $\Omega$. For $r \in (|\Omega|, \infty)$ let $\Omega_r$ be the element of $\Omega$ whose radius is $r$. If $(\alpha, \beta) \in K^2$, then for sufficiently small $r \in (|\Omega|, \infty)$ the function $\lambda \mapsto |\alpha - \beta\lambda|$ is constant on $\Omega_r$; let $\nu_\Omega(\alpha, \beta)$ be its value.

It turns out that the map $\Omega \mapsto \nu_\Omega$ establishes a bijective correspondence between the equivalence classes of norms and the equivalence classes of holes: every $\nu_\Omega$ lies in

$\mathcal{N}$, for every $\nu \in \mathcal{N}$ there is a hole $\Omega$ with $\nu \sim \nu_\Omega$, and for any two holes, $\Omega$ and $\Omega'$, one has

$$\nu_\Omega \sim \nu_{\Omega'} \iff \Omega \sim \Omega'.$$

Thus, our problem may also be formulated as: ***describe the set of all equivalence classes of the holes.***

## 3. MULTI-ORTHOGONAL BASES

The following observation was made already in 1964 by Carpentier [2]. (It also follows easily from [6], Theorem 1.11).

Let $\| \ \|_1, \| \ \|_2$ be norms on a finite-dimensional space $E$ such that $(E, \| \ \|_1)$, $(E, \| \ \|_2)$ both have orthogonal bases. Then there exists a base of $E$ which is orthogonal to both $\| \ \|_1$ and $\| \ \|_2$.

**Problem.** *Do we have a similar result for three (finitely many) norms? For which $K$ is the 'natural' infinite dimensional version of the above true?*

## 4. CARTESIAN SPACES

In the terminology of S. Bosch *et al.* [1], a normed vector space is *Cartesian* if every finite dimensional subspace has an orthogonal base.

Let us call a normed space ***Hilbertian*** of every 1-dimensional linear subspace has an orthogonal complement. Every finite dimensional linear subspace of a Hilbertian space has an orthogonal complement; it follows that a Hilbertian space is Cartesian.

**Problem.** *Is every Cartesian normed space Hilbertian?*

If $K$ is spherically complete, then every normed space is Hilbertian and Cartesian. In general, every normed space that has an orthogonal base is Hilbertian and Cartesian. Conversely, a Cartesian space of countable type has an orthogonal base, hence is Hilbertian.

However, if the valuation of $K$ is dense, there exists a Cartesian space without orthogonal base. We sketch a proof of this statement. Let $\check{c}_0$ be the spherical completion of $c_0$. By Zorn's Lemma, there is a maximal Cartesian subspace $E$ of $\check{c}_0$ containing $c_0$. Suppose $E$ has an orthogonal base. Then $E$ has a countable orthonormal base $\{a_1, a_2, \ldots\}$. Choose $\varepsilon_1, \varepsilon_2, \ldots \in K$ with $|\varepsilon_1| > |\varepsilon_2| > \ldots \to 1$. Take $u \in \check{c}_0$ such that

$$\|u - \sum_{n=1}^{N} \varepsilon_n a_n\| \leq |\varepsilon_{N+1}| \qquad (N = 1, 2, \ldots).$$

Then $u \notin E$. Set $F = E + [u]$. Putting

$$b_N = u - \sum_{n=1}^{N} \varepsilon_n a_n \qquad (N = 1, 2, \ldots)$$

one can show that $\{u, b_1, b_2, \ldots\}$ is an orthogonal base for $F$, so that $F$ is Cartesian. This contradicts the maximality of $E$.)

## 5. ORTHOGONAL ALMOST COMPLEMENTS

A closed subspace $S$ of $E$ is said to be an *orthogonal almost complement* (o.a.c.) of a closed subspace $T$ if $S \perp T$ and $S + T$ has finite codimension in $E$. (To show that this concept does not appear out of the blue consider the following easily proved **Proposition.** *The 'open' balls $B(0, r^-)$ are weakly closed for each $r > 0$ if and only if each onedimensional subspace has an o.a.c.)*

**Problem.** (Compare [5], Lemma 4.35 (iii).) *If each onedimensional subspace has an o.a.c. then does it follow that each finite dimensional subspace has an o.a.c.?*

## 6. IMAGES OF COMPACTOIDS

If $A \subset E$ is a compactoid and $T \in \mathcal{L}(E, F)$ is injective then $T$ maps $A$ homeomorphically onto $TA$. This leads to the

**Problem.*** *Let $A \subset E$ be an absolutely convex compactoid and let $T \in \mathcal{L}(E, F)$. If*

* *Note of the editors:* A solution to this problem was obtained during the production of this book; it appears in this same volume as Appendix B.

*$y_1, y_2, \ldots$ is a sequence in $TA$ with $\lim_{n\to\infty} y_n = 0$ does there exist a sequence $x_1, x_2, \ldots$ in some scalar multiple of $A$ with $\lim_{n\to\infty} x_n = 0$ and $Tx_n = y_n$ for each $n$?*

(It suffices to consider the case where $T$ is a quotient map. If in the above we require the $x_1, x_2, \ldots$ to be in $A$ the problem has a negative answer. In fact, a positive answer would imply that $TA$ is closed whenever $A$ is closed, which is not true if $K$ is not spherically complete; see [6], Theorem 6.28.)

## 7. STRONG POLARITY

(See [4] for definitions and for related concepts and problems.) It is well known that $E$ is strongly polar (SP) if and only if each closed edged absolutely convex subset of $E$ is polar. Let us define:

$E$ is *boundedly strongly polar* (BSP) if every bounded closed edged absolutely convex subset of $E$ is polar or, equivalently, if each norm inducing the topology is polar.

$E$ has the *almost orthogonal complementation property* (AOCP) if, for each $t \in (0,1)$, every closed linear subspace of $E$ has a $t$-orthogonal complement.

Obviously we have
$E$ is of countable type $\Longrightarrow$ $E$ has AOCP $\Longrightarrow$ $E$ is SP $\Longrightarrow$ $E$ is BSP.

**Problem.** *Let $K$ be not spherically complete. Which ones of the opposite implications are true?*

## 8. REFLEXIVITY

**Problem.** *Let $a \in E$, $a \neq 0$. If $E/[a]$ is reflexive, must $E$ itself be reflexive?*

(If $E$ is reflexive, then so is $E/[a]$. Proof. Let $\varepsilon > 0$; we are done if we can find a reflexive space $D$ and a linear bijection $T : D \to E/[a]$ with $\|T\| \leq 1$, $\|T^{-1}\| \leq 1 + \varepsilon$. There exists an $f \in E'$ such that $f(a) = 1$, $\|f\|\ \|a\| \leq 1 + \varepsilon$. Take $D = f^{-1}(0)$ and let $T$ be the restriction of the quotient map $E \to E/[a]$.)

A Banach space $F$ is called *pseudoreflexive* (or *polar*) if the natural map

$$j_F : F \to F''$$

is isometric. The following variation on the above problem is just as meaningful: *if $E/[a]$ is pseudoreflexive, must $E$ be pseudoreflexive too?*

However, there exist a Banach space $E$ and an $a \in E$, $a \neq 0$ such that $j_{E/[a]}$ is injective but $j_E$ is not! Such $E$ and $a$ can be made as follows. For every $n \in \mathbf{N}$ let $F_n$ be a Banach space such that $F_n'$ separates the points of $F_n$ and there exists an $a_n \in F_n$ with

$$\|a_n\| = 1, \quad |f(a_n)| \leq n^{-1}\|f\| \qquad (f \in E_n').$$

(See [5], 3.I and 4.N; take $a_n = (1, 1, 1, \ldots)$.) Let $F = \oplus F_n$, $D = \{(\lambda_1 a_1, \lambda_2 a_2, \ldots) : \lambda_1, \lambda_2, \ldots \in K, \sum \lambda_n = 0\}$, $E = F/D$ and let $a \in E$ be the element corresponding to $(a_1, 0, 0, \ldots) \in F$.

A related problem is:

**Problem.** *Let $E$ be a Banach space. Let $A, B$ be closed linear subspaces with $A + B = E$, $A \cap B = \{0\}$ and suppose both are (pseudo)reflexive. Does it follow that $E$ is (pseudo)reflexive?*

## 0. WEAKLY CLOSED CONVEX SUBSETS OF $c_0$

If $K$ is spherically complete each closed absolutely convex subset of $c_0$ is weakly closed. In general, each closed *edged* absolutely convex subset of $c_0$ is weakly closed. Also, closed compactoids are weakly closed. However, if $K$ is not spherically complete, one can always find a closed absolutely convex subset of $c_0$ which is not weakly closed ([8], Theorem 1.1).

**Problem.** *Characterize the weakly closed absolutely convex subsets of $c_0$.*

See also Problem 5.

## 10. CLOSED CONVEX SETS DETERMINING TOPOLOGY

Let $K$ be spherically complete and $\dim E = \infty$. Then the norm topology and the weak topology differ, yet these topologies have the same collection of closed convex sets. On the other hand, if $K$ is not spherically complete and $\dim E = \infty$ it follows from [8], Theorem 1.1 that one can find a norm closed absolutely convex set that is not weakly closed. This leads naturally to the following general

**Problem.** *Let $\tau_1, \tau_2$ be two locally convex topologies on a vector space over a non-spherically complete $K$. Suppose $\tau_1$-closed = $\tau_2$-closed for absolutely convex sets. Does it follow that $\tau_1 = \tau_2$?*

## 11. HYPERISOMORPHIC SPACES

Let us say that $E$ is *hyperisomorphic* if there exists a closed hyperplane which is linearly homeomorphic to $E$.

**Problem.** *Is, for spherically complete $K$, every infinite dimensional Banach space hyperisomorphic? More generally, is every polar infinite dimensional Banach space hyperisomorphic?*

## 12. COMPLETELY CONTINUOUS OPERATORS

Let us say that an $A \in \mathcal{L}(E, F)$ is *completely continuous* if for every sequence $x_1, x_2, \ldots$ in $E$

$$\lim_{n\to\infty} x_n = 0 \text{ weakly} \implies \lim_{n\to\infty} \|Ax_n\| = 0.$$

Clearly every compact operator is completely continuous. The converse is not true as for spherically complete $K$ *every* $A \in \mathcal{L}(E, F)$ is completely continuous.

**Problem.** *For nonspherically complete $K$, characterize the completely continuous operators in $\mathcal{L}(E, F)$.*

## 13. THE INVARIANT SUBSPACE PROBLEM

If $T \in \mathcal{L}(E)$, must $E$ have a nontrivial closed linear subspace that is invariant under $T$?

If $K$ is not algebraically closed, the answer clearly is negative: Let $E \neq K$ be an algebraic field extension of $K$ generated by an element $a$ and let $Tx = ax$ $(x \in E)$.

The answer is also negative if $K$ is not spherically complete. To see this, provide the spherical completion $\check{K}$ of $K$ with the structure of a valued field ([5], Th. 4.49), take $a \in \check{K}\backslash K$, let $E$ be the closed linear hull of $\{1, a, a^2, \ldots\}$ in $\check{K}$ and $Tx = ax$ $(x \in E)$. Then $E$ is a subalgebra of $\check{K}$. Actually, $E$ is a field: If $x \in E\backslash\{0\}$, there exists a $\xi \in K$ with $|x - \xi| < |\xi|$; then

$$x^{-1} = \xi^{-1} \sum_{n=0}^{\infty} (1 - \xi^{-1}x)^n \in E.$$

Every closed linear subspace of $E$ that is invariant under $T$ is an ideal in $E$, hence must be either $E$ or $\{0\}$.

There remains the following

**Problem.** *Assume that $K$ is algebraically closed and spherically complete. Let $T \in \mathcal{L}(E)$. Does $E$ necessarily have a nontrivial closed linear subspace that is invariant under $T$?*

If we restrict ourselves to infinite dimensional Banach spaces, the reasoning we gave for non-algebraically closed $K$ falls through and the one for non-spherically complete $K$ works only if $\check{K}$ contains an element that is not algebraic over $K$. This brings us to a question that in itself has nothing to do with invariant subspaces:

**Problem.** *If $K$ is not spherically complete, can $\check{K}$ consist only of elements that are algebraic over $K$?*

## 14. ARCHIMEDEAN NORMS

An *A-norm* ([5]) on a vector space $D$ is a function $q : D \to [0, \infty)$ satisfying

$$q(\lambda x) = |\lambda| q(x) \qquad (\lambda \in K,\ x \in D),$$

$$q(x + y) \leq q(x) + q(y) \qquad (x, y \in D),$$

$$q(x) = 0 \implies x = 0 \qquad (x \in D).$$

Such an $A$-norm induces a vector space topology.

**Problem.** (See [4]) *Let $q : D \to [0, \infty)$ be an $A$-norm and assume that for every linear subspace $D_0$ of $D$, every continuous linear function $D_0 \to K$ extends to a continuous linear function $D \to K$. Must there exist a (non-Archimedean) norm $\| \quad \|$ on $E$ that is equivalent to $q$?*

(There is a natural candidate for $\| \quad \|$:

On the space $D'$ of all $q$-continuous linear functions $D \to K$ we impose a (non-Archimedean) norm $q'$ by

$$q'(f) = \inf\{c \in [0, \infty) : |f(x)| \leq cq(x) \text{ for all } x \in D\}.$$

$q'$ determines a non-Archimedean norm $\overline{q}$ on $D$:

$$\overline{q}(x) = \inf\{c \in [0, \infty) : |f(x)| \leq cq'(f) \text{ for all } f \in D'\}.$$

Then $\overline{q} \leq q$ and $D'$ is just the space of all $\overline{q}$-continuous linear functions on $D$.

For $x_1, x_2, \ldots \in D$ write

$$x_n \rightharpoonup 0$$

if $f(x_n) \to 0$ for all $f \in D'$. Trivially,

$$q(x_n) \to 0 \implies x_n \rightharpoonup 0.$$

By 5.2 of [6],

$$x_n \rightharpoonup 0 \iff \overline{q}(x_n) \to 0$$

and

$$x_n \rightharpoonup 0 \iff \|x_n\| \to 0$$

if $\| \quad \|$ is a norm providing a positive answer to our Problem. It follows that, if such a norm $\| \quad \|$ exists at all, then $\overline{q}$ is one. Also, this is the case if and only if

$$x_n \rightharpoonup 0 \implies q(x_n) \to 0.$$

Consequently, to solve the Problem one may restrict oneself to spaces $D$ of countable type.

Van Gisbergen ([3]) gives an example of an $A$-norm that does not have the extension property mentioned in the Problem although $D'$ separates the points of $D$.)

## 15. ULTRAMETRIZABILITY

Let $\ell^1$ be the space of all sequences $x = (x_1, x_2, \ldots)$ where $x_i \in K$ for each $i$ such that $\|x\| := \sum_i |x_i| < \infty$. Then $\ell^1$ is a $K$-Banach space with respect to the $A$-norm $\| \quad \|$ (see Problem 14).

**Problem.*** *Is $\ell^1$ ultrametrizable?*

(One can show that the dense subspace $c_{00} := \{(x_1, x_2, \ldots) : x_i = 0 \text{ for large } i\}$ is ultrametrizable. For separable $K$ this is easily seen as follows. One verifies that $B(0, r^-)$ is closed in $c_{00}$. Then $c_{00}$ is zero-dimensional, separable and metrizable, hence ultrametrizable.)

An affirmative answer would solve the more general

**Problem.** *Does there exist a complete ultrametrizable topological vector space over $K$ which is not locally convex?*

## REFERENCES

[1] S. Bosch, U. Güntzer and R. Remmert. *Non-Archimedean Analysis.* Springer-Verlag, Berlin-Heidelberg-New York-Tokyo, 1984.

[2] J.-P. Carpentier. *Semi-normes et ensembles convexes dans un espace vectoriel sur un corps ultramétrique.* Sém. Choquet (1964/65), No. 7.

[3] J. van Gisbergen. *Archimedisch genormeerde vectorruimten over een niet-Archimedisch gewaardeerd lichaam.* Master's Thesis, Department of Mathematics, Catholic University, Nijmegen, The Netherlands, 1985.

* *Note of the editors:* A solution to this problem was obtained during the production of this book; it appears in this same volume as Appendix A.

[4] C. Pérez-García. *The Hahn-Banach extension property in p-adic analysis.* In *p-adic Functional Analysis*, edited by J.M. Bayod, N. De Grande-De Kimpe and J. Martínez-Maurica, Marcel Dekker, New York (1991), 127-140.

[5] A. van Rooij. *Non-Archimedean Functional Analysis.* Marcel Dekker Inc., New York-Basel, 1978.

[6] A. van Rooij. *Notes on p-adic Banach spaces.* Reports 7633 and 7725, Dept. Math., Catholic University, Nijmegen, The Netherlands (1976, 1977).

[7] W.H. Schikhof. *The continuous linear image of a p-adic compactoid.* Nederl. Akad. Wetensch. Indag. Math. **51** (1989), 119–123.

[8] W.H. Schikhof. *The p-adic Krein-Šmulian theorem.* In *p-adic Functional Analysis*, edited by J.M. Bayod, N. De Grande-De Kimpe and J. Martínez-Maurica, Marcel Dekker, New York (1991), 177–189.

KATHOLIEKE UNIVERSITEIT
MATHEMATISCH INSTITUUT
TOERNOOIVELD
6525 ED NIJMEGEN, THE NETHERLANDS

# APPENDIX A

## THE SPACE $\ell^1(K)$ IS NOT ULTRAMETRIZABLE

José M. Bayod*

This note is devoted to prove the statement in the title —whenever the valuation of $K$ is not trivial— thus answering a question posed by A.C.M. van Rooij and W.H. Schikhof in [5], p. 218.

**1.** Let $K$ be a field endowed with a nonarchimedean valuation, and let $\ell^1(K)$, or $\ell^1$ for short, be the space of all sequences $x = (x_i)_{i\in \mathbb{N}}$ of members of $K$ such that $\|x\|_1 := \sum |x_i|$ is finite; then $\| \ \|_1$ is an archimedean norm (or *A-norm*, in the teminology of [4]; see also [5], p. 216) on $\ell^1$ that makes it into a $K$-Banach space.

If the valuation of $K$ is trivial (that is to say, if $|t| = 1$ for every $t \in K\backslash\{0\}$), then the set $\ell^1$ is the same as the set $c_{00}$ of eventually null sequences over $K$, and the topology defined by $\| \ \|_1$ is the discrete topology (any open ball of radius 1 is a singleton), then it is the topology associated to the trivial ultrametric.

IN THE SEQUEL, WE ASSUME THAT THE VALUATION OF $K$ IS NOT TRIVIAL.

**2.** As a matter of fact, even when $K$ is non-trivially valued, the space $c_{00}$ (which is dense in $\ell^1$) can be shown to be ultrametrizable. This is proved for the "$K$ separable" case, and stated for the general case, in [5], p. 218; a proof valid for all cases can be sketched as follows:

* Research partially supported by DGICYT (grant PS87-0094)

- the space $c_{00}$ is the union of its closed subspaces $K^m$, $m \in \mathbb{N}$, and all these $K^m$ are ultrametrizable in a natural way;
- if an ultrametrizable space has a countable closed cover of strongly zero-dimensional spaces, then it is strongly zero-dimensional. (Cf. [2], Th. 7.2.1);
- for a nonempty metrizable space, being strongly zero-dimensional is equivalent to being ultrametrizable (see, e.g., [2], Pr. 7.3.F).

**3.** Then it seems natural to ask if the ultrametrizability property of $c_{00}$ carries over up to the whole $\ell^1$. The answer is no: we prove that $\ell^1$ is not even weakly zero-dimensional.

In some sense, the situtation can be regarded as analogous to that of the common spaces $\mathbb{Q}$ and $\mathbb{R}$, where $\mathbb{Q}$ is dense in $\mathbb{R}$, $\mathbb{Q}$ is ultrametrizable, and $\mathbb{R}$ is not; nevertheless, the analogy is far from complete, the position of $\ell^1$ being much more delicate (as long as connectendness properties are concerned) than that of $\mathbb{R}$: $\mathbb{R}$ is connected and $\ell^1$ is totally disconnected, that is, its quasi-components are its singletons (see point **5** below).

THEOREM. *If the valuation of $K$ is not trivial, then the space $\ell^1(K)$ is not zero-dimensional.*

*Proof.* Recall that a topological space is called zero-dimensional if it is a non-empty $T_1$-space and has a base consisting of closed (and open) sets. Consider an open neighborhood $U$ of the null sequence 0 which is contained in the unit ball,

$$U \subset \{(x_i) \in \ell^1 : \sum |x_i| < 1\}\,;$$

we shall prove that its topological boundary $\operatorname{Fr} U$ is not empty; when we have proved this, it will be obvious that $\ell^1$ is not zero-dimensional, since the point 0 cannot have a fundamental system of closed and open neighborhoods: it has none within the unit ball.

For $n_1, \ldots, n_k \in \mathbb{N}$, call

$$x(n_1, \ldots, n_k) := (t^{n_1}, \ldots, t^{n_k}, 0, 0, \ldots) \in \ell^1,$$

where $t \in K$ is a fixed member of $K$ with $0 < |t| < 1$, and define inductively a sequence in N, $(m_k)_k$, as follows:

$$\begin{aligned} m_1 &:= \min\{n \in \mathbf{N} : x(n) \in U\} \\ &\cdots \\ m_k &:= \min\{n \in \mathbf{N} : x(m_1, \ldots, m_{k-1}, n) \in U\} \\ &\cdots \end{aligned}$$

(it is easy to check that these mimima exist). Finally, define $a := (t^{m_1}, \ldots, t^{m_k}, \ldots)$. Then

- $a \in \ell^1$:

$$\forall k, \quad |t^{m_1}| + \ldots + |t^{m_k}| = \|x(m_1, \ldots, m_k)\|_1 < 1,$$

  so $\sum_{k=1}^{\infty} |t^{m_k}| \leq 1$, and therefore $a \in \ell^1$.

- $a \in \overline{U}$:

$$\|a - x(m_1, \ldots, m_k)\|_1 = \sum_{i>k} |t^{m_i}| \to 0 \quad (k \to \infty).$$

- $a \in \overline{\ell^1 \backslash U}$:

  for those $m_k$'s that are greater than 1 (almost all, since $\sum_{k=1}^{\infty} |t^{m_k}| \leq 1$),

$$\|a - x(m_1, \ldots, m_{k-1}, m_k - 1)\|_1 = \sum_{i>k} |t^{m_i}| + |t^{m_k - 1}||t - 1| \to 0 \quad (k \to \infty).$$

Hence, $a \in \operatorname{Fr} U$.

**4.** The fact that the metric induced by $\| \ \|_1$ on $\ell^1$ has no topologically equivalent ultrametric has another interesting consequence:

On $\ell^1$ we can also consider the ultrametric associated to the uniform norm,

$$\|x\|_\infty := \max |x_n|, \quad x = (x_n) \in \ell^1.$$

It is obvious that $\| \ \|_\infty \leq \| \ \|_1$, so that the natural metric on $\ell^1$ has a subdominant ultrametric, in the terminology of [1]; but now we have proved that the metric space $\ell^1$ is *not* ultrametrizable, thus providing a negative answer to a question raised in [1], p. 831. (The first example that settled this question was proposed by J. Kulesza, [3].)

**5.** On the other hand, from the observation that $\ell^1$ has a subdominant ultrametric it is easy to deduce that it is totally disconnected in the sense of [2], i.e., that for each point $x$ in $\ell^1$, the intersection of all closed and open subsets of $x$ which contain $x$ is the singleton $\{x\}$ (this property is stronger, in general, than the property of having the singletons as the only connected nonempty subsets).

This is in sharp contrast with properties of the apparently similar A-normed space $L^1([0,1])$: if $E$ is the set of all Borel measurable functions from the real interval $[0,1]$ into $K$ for which the Lebesgue integral $\int_0^1 |f(x)|\,dx$ is finite, and

$$N := \{f \in E : f = 0 \text{ almost everywhere}\},$$

then on the quotient space $L^1([0,1]) := E/N$ we can consider an A-norm defined by

$$\|\text{class of } f\|_1 := \int_0^1 |f(x)|\,dx.$$

This space, much unlike $\ell^1$, is even arcwise connected! (see [4], p. 89).

Also, one should be careful not to draw the wrong conclusion from the property stated as an intermediate step in the proof of the theorem: if $Y$ is the open unit ball of $\ell^1$, then $Y$ must be totally disconnected, for the same reason as $\ell^1$; therefore the point $0 \in Y$ necessarily has plenty of closed and open neighborhoods *as a member of the subspace* $Y$, although none of them is closed in $\ell^1$.

**6.** All results stated and proved in this note can also be given (with minor modifications) for each space $\ell^q(K)$ ($1 \leq q < \infty$), of $q$-summable sequences in $K$, endowed with the A-norm $\|x\|_q := (\sum_i |x_i|^q)^{1/q}$.

The author is happy to acknowledge conversations with W.H. Schikhof that improved the presentation of this paper.

## REFERENCES

[1] José M. Bayod and J. Martínez-Maurica. *Subdominant ultrametrics*. Proc. Amer. Math. Soc. **109** (1990), 829–834.

[2] R. Engelking. *General topology*. Heldermann Verlag, Berlin, 1989.

[3] J. Kulesza. Personal communication. 1990.

[4] A.C.M. van Rooij, *Non-archimedean Functional Analysis*, Marcel Dekker, New York, 1978.

[5] A.C.M. van Rooij and W.H. Schikhof. *Open problems*. In *p-adic Functional Analysis*, edited by J.M. Bayod, N. De Grande-De Kimpe and J. Martínez-Maurica, Marcel Dekker, New York (1991), 209–219.

DEPARTAMENTO DE MATEMÁTICAS
FACULTAD DE CIENCIAS, UNIVERSIDAD DE CANTABRIA
AV. DE LOS CASTROS
39071 SANTANDER, SPAIN

# APPENDIX B

## ZERO SEQUENCES IN $p$-ADIC COMPACTOIDS

W.H. Schikhof

**ABSTRACT.** We prove the following theorem, solving Problem 6 of [3]. *Let* $E, F$ *be Banach spaces over a nonarchimedean valued field* $K$, *let* $A$ *be an absolutely convex closed compactoid in* $E$ *and* $T \in \mathcal{L}(E, F)$. *Then, if* $y_1, y_2, \ldots$ *is a sequence in* $TA$ *tending to* 0 *then there is a sequence* $x_1, x_2, \ldots$ *in some scalar multiple of* $A$ *tending to* 0 *such that* $Tx_n = y_n$ *for each* $n$. (See Theorem 2.7).

**INTRODUCTION.** For a proof we decompose $T$:

$$\begin{array}{ccc} A & \xrightarrow{T} & TA \\ \pi \searrow & & \nearrow i \\ & A/A \cap \operatorname{Ker} T & \end{array}$$

Here, the topology of $TA$ is inherited from the norm on $F$ and on $A/A \cap \operatorname{Ker} T$ we take the quotient topology induced by the quotient map $\pi$. Then $i$ (which is the unique map making the diagram commute) is continuous. The object $A/A \cap \operatorname{Ker} T$ is in a natural way a topological module over $\{\lambda \in K : |\lambda| \leq 1\}$ whose topology is induced by the metric $(\pi(x), \pi(y)) \mapsto \|\pi(x) - \pi(y)\|$ where

$$\|\pi(z)\| = \operatorname{dist}(z, A \cap \operatorname{Ker} T) \qquad (z \in A)$$

This "norm" $\| \ \|$ on $A/A \cap \operatorname{Ker} T$ satisfies the strong triangle inequality and

$$(*) \qquad \|\lambda \pi(z)\| \leq |\lambda| \, \|\pi(z)\| \qquad (z \in A, \lambda \in K, |\lambda| \leq 1)$$

but one does not always have equality in $(*)$. Thus, we shall study a class of tolopogical modules over the valuation ring which properly contains the class of absolutely convex subsets of locally convex spaces. This study involves a careful modification of the theory given in [4] for absolutely convex sets. We then will be able to conclude that although $i$ (see diagram) is not always a homeomorphism it does have the property that $i(t_n) \to 0$ implies $\|\lambda t_n\| \to 0$ for each $\lambda \in K$, $|\lambda| < 1$; the statement announced in the Abstract follows easily.

**TERMINOLOGY.** (For unexplained terms we refer to [2])

1. Throughout $K$ is a nonarchimedean nontrivially valued field, that is complete under the metric induced by the valuation $|\ |$. We set

$$B_K := \{\lambda \in K : |\lambda| \leq 1\}$$
$$B_K^- := \{\lambda \in K : |\lambda| < 1\}.$$

2. Let $A$ be a module over the ring $B_K$, let $B$ be a submodule of $A$, let $\lambda, \mu \in K$. We set

$$\lambda B := \begin{cases} \{\lambda b : b \in B\} & \text{if } |\lambda| \leq 1 \\ \{x \in A : \lambda^{-1}x \in B\} & \text{if } |\lambda| \geq 1 \end{cases}$$

(observe that this causes no ambiguity if $|\lambda| = 1$). We have the following obvious consequences.

(i) $\lambda B$ is a submodule of $A$, $1 \cdot B = B$.
(ii) If $|\lambda| \leq |\mu|$ then $\lambda B \subset \mu B$. In particular, $|\lambda| = |\mu|$ implies $\lambda B = \mu B$.
(iii) If either $|\lambda|, |\mu| \leq 1$ or $|\lambda|, |\mu| \geq 1$ then $(\lambda\mu)B = \lambda(\mu B) = \mu(\lambda B)$.

We shall say that $B$ *absorbs* a subset $X$ of $A$ if $X \subset \bigcup_{\lambda \in K} \lambda B$. The module generated by $X \subset A$ is denoted $\operatorname{co} X$.

3. A *locally convex topology* on a $B_K$-module $A$ is a topology $\tau$ on $A$ such that

(i) $(A, \tau)$ is a topological $B_K$-module (of course, the topology on $B_K$ is the valuation topology),
(ii) there is a neighbourhood base of 0 consisting of ($\tau$-open) $B_K$-submodules of $A$.

Then we call $A = (A, \tau)$ a *locally convex module.*

Clearly, absolutely convex subsets of locally convex spaces over $K$ are examples of locally convex modules. Any submodule of a locally convex module is, with the restriction topology, a locally convex module. If $B$ is a (closed) submodule of a locally convex module $A$ the quotient topology on $A/B$ is (Hausdorff) and locally convex. Let $A$ be a locally convex module. The closure of a subset $X$ of $A$ is denoted $\overline{X}$. Instead of $\overline{\text{co}X}$ we write $\overline{\text{co}}X$. Let $X_1, X_2, \ldots$ be subsets of $A$. We shall write $\lim_{n\to\infty} X_n = 0$ if for each neighbourhood $U$ of 0 in $A$ we have $X_n \subset U$ for large $n$. The following is not hard to see. If $X_1, X_2, \ldots$ are submodules, $\lim_{n\to\infty} X_n = 0$ then

$$\bigcap_{n=1}^{\infty} (\overline{B + X_n}) = \overline{B}$$

for every submodule $B$ of $A$.

Let $i \mapsto x_i$ be a net in a locally convex module converging to 0. Then, for any net $i \mapsto \lambda_i$ in $B_K$, the net $i \mapsto \lambda_i x_i$ converges to 0. This is a direct consequence of local convexity.

THROUGHOUT §1 AND §2 WE ASSUME THAT THE VALUATION OF $K$ IS DENSE.

## §1. LOCALLY CONVEX $B_K$-MODULES

In this section $A$ is a locally convex module over $B_K$.

LEMMA 1.1. *Let $B$ be a closed submodule of $A$ let $a \in A$, let $\lambda \in B_K^-$. If*

$$i \mapsto x_i = b_i + c_i a \qquad (b_i \in B,\ c_i \in B_K)$$

*is a net in $B + \text{co}\{a\}$ converging to 0 then $\lim_i c_i = 0$ or $\lambda x_i \in B$ eventually, where the latter case occurs if $B$ absorbs $\{a\}$.*

*Proof.* Set $C := \{\xi \in B_K : \xi a \in B\}$. Then $C$ is absolutely convex; let $r$ be its diameter. Suppose $\overline{\lim_i}|c_i| > r$. Then there exists a $\mu \in K$ and a cofinal $J \subset I$ such that

$$|c_j| \geq |\mu| > r \qquad (j \in J)$$

Then $\mu c_j^{-1} \in B_K$ so that

$$\mu c_j^{-1} b_j + \mu a = \mu c_j^{-1} x_j \to 0$$

We see that $\mu a = -\lim_j \mu c_j^{-1} b_j \in \overline{B} = B$ so that $\mu \in C$ conflicting $|\mu| > r$. Thus, we have proved that $\overline{\lim_i}|c_i| \leq r$. If $r = 0$ then $\lim_i c_i = 0$. If $r > 0$ then, eventually, $|\lambda c_i| < r$, i.e. $\lambda c_i \in C$ i.e. $\lambda c_i a \in B$ i.e. $\lambda x_i = \lambda b_i + \lambda c_i a \in \lambda B + B \subset B$.
To prove the second assertion, let $B$ absorb $\{a\}$ and let $\lim_i c_i = 0$. We have $\mu a \in B$ for some nonzero $\mu \in B_K$. Then, eventually, $|c_i| \leq |\mu|$ yielding $c_i a \in B$ implying $x_i \in B$.

For a submodule $B$ of $A$ we define

$$B^e := \bigcap \{\lambda B : \lambda \in K, |\lambda| > 1\}$$

The following elementary facts are easily verified. *$B^e$ is a submodule, $B \subset B^e$, $B^{ee} = B^e$, $B^e = \{x \in A : \lambda x \in B$ for all $\lambda \in B_K^-\}$. $\overline{B^e} \subset \overline{B}^e$.* Further, we have: *$B^e$ is closed $\Longleftrightarrow \overline{B} \subset B^e \Longleftrightarrow \lambda \overline{B} \subset B$ for each $\lambda \in B_K^-$.*

**LEMMA 1.2.** *Let $B$ be a closed submodule of $A$, let $a \in A$. Then $(B+\mathrm{co}\{a\})^e$ is closed.*

Proof. Let $i \mapsto x_i = b_i + c_i a$ ($b_i \in B, c_i \in B_K$) be a net in $B + \mathrm{co}\{a\}$ converging to some $x \in A$. Let $\lambda \in B_K^-$; we shall prove that $\lambda x \in B + \mathrm{co}\{a\}$. In fact, the net $(i,j) \mapsto x_i - x_j$ converges to 0. So by Lemma 1.1 we have either $\lim_{i,j}(c_i - c_j) = 0$ (then $\lim_i c_i = c$ for some $c \in B_K$ and it follows easily that even $x \in B + \mathrm{co}\{a\}$) or $\lambda(x_i - x_j) \in B$ for, say, all $i, j \geq s$. Then $\lambda x_i \in \lambda x_s + B$ ($i \geq s$) so $\lambda x = \lim_i \lambda x_i \in \overline{\lambda x_s + B} = \lambda x_s + B \in \lambda(B + \mathrm{co}\{a\}) + B \subset B + \mathrm{co}\{a\}$.

**LEMMA 1.3.** *Let $B$ be a closed submodule of $A$, let $a_1, \ldots, a_n \in A$. Then $(B + \mathrm{co}\{a_1, \ldots, a_n\})^e$ is closed.*

Proof. Lemma 1.2 covers the case $n=1$. Suppose, for some $m$,
$Z := (B + \mathrm{co}\{a_1, \ldots, a_{m-1}\})^e$ is closed. Then, again by Lemma 1.2,
$\overline{B + \mathrm{co}\{a_1, \ldots, a_m\}} \subset \overline{Z + \mathrm{co}\{a_m\}} \subset (Z + \mathrm{co}\{a_m\})^e = (B + \mathrm{co}\{a_1, \ldots, a_m\})^e$, so $(B + \mathrm{co}\{a_1, \ldots, a_m\})^e$ is closed.

LEMMA 1.4. *Let $B$ be a closed submodule of $A$, let $a_1, \ldots, a_n \in A$, and suppose that $B$ absorbs $\{a_1, \ldots, a_n\}$. If $i \mapsto x_i$ is a net in $B + \text{co}\{a_1, \ldots, a_n\}$ converging to 0 then $\lambda x_i \in B$ eventually, for each $\lambda \in B_K^-$.*

*Proof.* Choose $\mu_1, \ldots, \mu_n \in B_K^-$ such that $|\prod_{i=1}^{n} \mu_i|^2 \geq |\lambda|$. We have $x_i \in (B+\text{co}\{a_1, \ldots, a_{n-1}\}) + \text{co}\{a_n\}$. By Lemma 1.1 we have, eventually,

$$\mu_1 x_i \in \overline{B + \text{co}\{a_1, \ldots, a_{n-1}\}} \subset (B + \text{co}\{a_1, \ldots, a_{n-1}\})^e \quad \text{(Lemma 1.3)},$$

so that, eventually,

$$\mu_1^2 x_i \in B + \text{co}\{a_1, \ldots, a_{n-1}\}$$

Inductively we find

$$\mu_1^2 \cdots \mu_n^2 x_i \in B \qquad \text{eventually}$$

implying $\lambda x_i \in B$ eventually.

## §2. COMPACTOID MODULES

DEFINITION 2.1. A locally convex module $A$ is a *compactoid module* if for each $\lambda \in B_K^-$ and each neighbourhood $U$ of 0 in $A$ there exists a finite set $F \subset A$ such that

$$\lambda A \subset U + \text{co}\, F$$

**Remark.** An absolutely convex subset of a locally convex space over $K$ is a compactoid module if and only if it is a compactoid in the usual sense. So Definition 2.1 generalizes the notion of compactoid to a larger class of objects, and we will see that it suits the purpose of the paper. (Yet we must warn the reader that, in general, compactoid modules are no longer 'compact-like': For any Banach space $E$ the module $\{x \in E : \|x\| \leq 1\}/\{x \in E : \|x\| < 1\}$ has the discrete topology but is a compactoid module.)

We need the following algebraic lemma.

LEMMA 2.2. *Let $B, U$ be submodules of a $B_K$-module $A$, let $x_1, \ldots, x_n \in A$ and suppose*

$$B \subset U + \text{co}\{x_1, \ldots, x_n\}.$$

*Then, for each $\lambda \in B_K^-$ there exist $b_1, \ldots, b_n \in B$ such that*

$$\lambda B \subset U + \mathrm{co}\{b_1, \ldots, b_n\}.$$

*Proof.* The proofs of Lemmas 1.1 and 1.2 of [1] apply in this more general situation.

**PROPOSITION 2.3.** *Every submodule of a compactoid module is compactoid.*

*Proof.* Let $B$ be a submodule of a compactoid module $A$, let $V$ be a zero neighbourhood in $B$, let $\lambda \in B_K^-$. Choose $\mu \in B_K^-$ with $|\lambda| \leq |\mu|^2$, choose an open submodule $U$ of $A$ with $U \cap B \subset V$. There is a finite set $F_1 \subset A$ such that

$$\mu B \subset \mu A \subset U + \mathrm{co}\, F_1$$

By Lemma 2.2 there exists a finite set $F_2 \subset B$ such that

$$\mu^2 B \subset U + \mathrm{co}\, F_2$$

It follows that

$$\lambda B \subset \mu^2 B \subset U \cap B + \mathrm{co}\, F_2 \subset V + \mathrm{co}\, F_2$$

which proves that $B$ is a compactoid module.

**PROPOSITION 2.4.** *Let $A$ be a metrizable compactoid module, let $\lambda \in B_K^-$. Then there exists a sequence $x_1, x_2, \ldots$ with $\lim_{n\to\infty} x_n = 0$ such that*

$$\overline{\lambda A} \subset \overline{\mathrm{co}}\{x_1, x_2, \ldots\} \subset A$$

*Proof.* Choose $\lambda_1, \lambda_2, \ldots \in B_K^-$ such that $|\Pi \lambda_n| \geq |\lambda|$. By metrizability and local convexity there exist open submodules

$$A = U_1 \supset U_2 \supset \cdots$$

of $A$ forming a neighbourhood base at 0. By compactoidity there is a finite set $F_1 \subset U_1$ such that

$$\lambda_1 U_1 \subset U_2 + \mathrm{co}\, F_1.$$

By Proposition 2.3 $U_2$ is a compactoid so

$$\lambda_2 U_2 \subset U_3 + \mathrm{co}\, F_2$$

for some finite set $F_2 \subset U_2$, etc. Inductively we find finite sets $F_1, F_2, \ldots$ such that

(i) $F_n \subset U_n$ for each $n$. Hence $\lim_n F_n = 0$ (see **TERMINOLOGY**, 3)

(ii) $\lambda_n U_n \subset U_{n+1} + \operatorname{co} F_n$ for each $n$.

Repeated application of (ii) yields for $n \in \mathbb{N}$

$$\begin{aligned}\lambda A &\subset \lambda_1\lambda_2\ldots\lambda_n A \subset \lambda_2\ldots\lambda_n(\lambda_1 U_1) \subset \lambda_2\ldots\lambda_n(U_2 + \operatorname{co} F_1)\\ &\subset \lambda_2\ldots\lambda_n U_2 + \operatorname{co} F_1 \subset \lambda_3\ldots\lambda_n U_3 + \operatorname{co} F_2 + \operatorname{co} F_1 \subset \cdots\\ &\subset U_{n+1} + \operatorname{co}(F_1 \cup F_2 \cup \ldots \cup F_n).\end{aligned}$$

We see that for each $n$

$$\overline{\lambda A} \subset U_{n+1} + \operatorname{co}\{F_1 \cup F_2 \cup \ldots\}$$

so that

$$\overline{\lambda A} \subset \overline{\operatorname{co}}\{F_1 \cup F_2 \cup \ldots\}$$

By enumerating $F_1, F_2, \ldots$ successively we obtain a sequence $x_1, x_2, \ldots$ in $A$ tending to 0 (as $\lim F_n = 0$) such that

$$\overline{\lambda A} \subset \overline{\operatorname{co}}\{x_1, x_2, \ldots\}.$$

**THEOREM 2.5.** *Let $E, F$ be locally convex $B_K$-modules, let $A$ be a complete metrizable compactoid submodule of $E$.*

(i) *If $F$ is Hausdorff and $T : E \to F$ is a continuous module homomorphism then $(TA)^e$ is closed.*

(ii) *If $B \subset E$ is a closed submodule of $E$ then $(A+B)^e$ is closed.*

*Proof.* We show that for a closed submodule $B$ of $F$ the set $(TA+B)^e$ is closed (then (i) follows by taking $B = \{0\}$, (ii) is the case $E = F$, $T$ is the identity). Thus, let $\lambda \in B_K^-$; we prove that $\lambda\overline{(TA+B)} \subset TA+B$.

Let $\lambda_1, \mu_1, \mu_2, \ldots \in B_K^-$ such that $|\lambda_1 \prod \mu_n| \geq |\lambda|$.

By Proposition 2.4 there exist $x_1, x_2, \ldots \in A$ such that

$$(*) \qquad \lim x_n = 0 \quad \text{and} \quad \overline{\lambda_1 A} \subset \overline{\operatorname{co}}\{x_1, x_2, \ldots\} \subset A$$

Write $X_n := \overline{\operatorname{co}}\{x_n\}$ and $A_n := \overline{\sum_{i \geq n} X_i}$. Then $\lim_{n\to\infty} X_n = 0$, $\lim_{n\to\infty} A_n = 0$ and $\lim_{n\to\infty} TA_n = 0$. First observe that by $(*)$

$$\lambda_1\overline{(TA+B)} \subset \overline{TA_1 + B}$$

Also we have by Lemma 1.2

$$\mu_1\overline{(TA_1+B)} = \mu_1\overline{(TX_1+TA_2+B)} \subset TX_1 + \overline{TA_2+B}$$
$$\mu_2\overline{(TA_2+B)} = \mu_2\overline{(TX_2+TA_3+B)} \subset TX_2 + \overline{TA_3+B}, \text{ etc.}$$

Let $x \in \lambda\overline{(TA+B)}$. Then $x \in (\prod \mu_n)\lambda_1\overline{(TA+B)} \subset TX_1 + (\prod_{n\geq 2} \mu_n)\overline{(TA_2+B)}$. There is a $y_1 \in X_1$ such that

$$x - Ty_1 \in \prod_{n\geq 2} \mu_n\overline{(TA_2+B)} \subset TX_2 + (\prod_{n\geq 3} \mu_n)\overline{(TA_3+B)}.$$

So there is a $y_2 \in X_2$ such that

$$x - Ty_1 - Ty_2 \in (\prod_{n\geq 3} \mu_n)\overline{(TA_3+B)}, \text{ etc.}$$

Inductively we find $y_1 \in X_1, y_2 \in X_2, \ldots$ such that for each $n$

$$x = T(y_1 + \cdots + y_n) + z_{n+1}$$

when $z_{n+1} \in \overline{TA_{n+1}+B}$. Now $\lim_{n\to\infty} y_n = 0$, so, by completeness of $A$, $y := \sum_{n=1}^{\infty} y_n$, exists, and lies in $A$. Then $z : \lim_{n\to\infty} z_{n+1}$ also must exist and it lies in $\bigcap_n \overline{(TA_n+B)}$ which is $B$ since $\lim_{n\to\infty} TA_n = 0$. We see that $x = Ty + z \in TA + B$.

As a corollary we obtain

**THEOREM 2.6.** *Let $A_1, A_2$ be locally convex modules, let $A_1$ be metrizable, complete and compactoid and let $T : A_1 \to A_2$ be a continuous injective module homomorphism. Let $i \mapsto a_i$ be a net in $A_1$ such that $Ta_i \to 0$. Then, for each $\lambda \in B_K^-$, $\lambda a_i \to 0$.*

*Proof.* Let $U$ be an open submodule of $A_1$, let $\lambda \in B_K^-$. Choose $\mu \in B_K^-$, $|\mu|^3 \geq |\lambda|$. There exists a finite set $F \subset A_1$ such that $\mu A \subset U + \operatorname{co} F$. Hence

$$\mu TA \subset TU + \operatorname{co} TF \subset (TU)^e + \operatorname{co} TF.$$

The net $i \mapsto \mu Ta_i$ lies in $(TU)^e + \operatorname{co} TF$ and converges to 0. Further, $(TU)^e$ is closed (Proposition 2.3 and Theorem 2.5) and it absorbs $TF$. Thus, by Lemma 1.4 we have, eventually: $\mu^2 Ta_i \in (TU)^e$ which implies $\mu^3 Ta_i \in TU$. As $|\lambda| \leq |\mu|^3$ we have $T(\lambda a_i) \in TU$. By injectivity of $T$, $\lambda a_i \in U$. If follows that $\lambda a_i \to 0$.

Now we can prove the result announced in the Abstract.

**THEOREM 2.7.** *Let $E, F$ be $K$-Banach spaces, let $T \in \mathcal{L}(E,F)$ and let $A \subset E$ be an absolutely convex and closed compactoid. If $y_1, y_2, \ldots$ is a sequence in $TA$ tending to 0 then for each $\lambda \in K$, $|\lambda| > 1$ there is a sequence $x_1, x_2, \ldots$ in $\lambda A$ tending to 0 such that $Tx_n = y_n$ for each $n$.*

*Proof.* Decompose $T$ (see the Introduction):

$$\begin{array}{ccc} A & \xrightarrow{T} & TA \\ \pi \searrow & & \nearrow i \\ & A/A \cap \operatorname{Ker} T & \end{array}$$

Let $z_n := i^{-1}(y_n)$ for each $n$, let $\lambda \in K$, $|\lambda| > 1$. By Theorem 2.6 $\lim_{n\to\infty} \lambda^{-1} z_n = 0$ in the quotient topology of $A/A \cap \operatorname{Ker} T$. By metrizability there exist $a_1, a_2, \ldots \in A$ such that $\lim_{n\to\infty} a_n = 0$ and $\pi(a_n) = \lambda^{-1} z_n$ for each $n$. Set $x_n := \lambda a_n$ ($n \in \mathbb{N}$). Then $x_n \in \lambda A$, $\lim_{n\to\infty} x_n = 0$ and $Tx_n = y_n$ for each $n$.

The following is a more general version of Theorem 2.7.

**THEOREM 2.8.** *Let $A_1, A_2$ be locally convex modules, let $A_1$ be metrizable, complete and compactoid and let $T : A_1 \to A_2$ be a continuous surjective module homomorphism. If $U \subset A_1$ is an open submodule then for every $\lambda \in K$, $|\lambda| > 1$ the module $T(\lambda U)$ is open in $A_2$.*

*Proof.* We may assume that $T$ is bijective (decompose $T$ in the spirit of above). If 0 is not in the interior of $T(\lambda U)$ then there exists a net $i \mapsto z_i$ in $A_2$ converging to 0 but $z_i \notin T(\lambda U)$ for each $i$. Let $x_i := T^{-1}(z_i)$. By Theorem 2.6 we have $\lambda^{-1} x_i \to 0$ so eventually $\lambda^{-1} x_i \in U$ i.e. $x_i \in \lambda U$ i.e. $z_i \in T(\lambda U)$, a contradiction.

## §3. THE CASE OF A DISCRETE VALUATION

We extend the definition of a compactoid module to arbitrary ground fields $K$ as follows. For $r \in [0,1]$ and a $B_K$-module $A$ set

$$rA := \bigcap \{\lambda A : \lambda \in K, |\lambda| \geq r\}$$

A locally convex module $A$ is a *compactoid* if for each $r \in [0,1)$ and each neighbourhood $U$ of 0 in $A$ there exists a finite set $F \subset A$ such that $rA \subset U + \operatorname{co} F$. If the valuation of $K$ is dense, this definition is equivalent to 2.1 whereas for discretely valued $K$ one obtains Definition 2.1 but where "$\lambda \in B_K^-$" is replaced by "$\lambda = 1$".

It is not hard to see that all lemmas, propositions and theorems of §1, §2 remain true for discretely valued $K$ even when we take $\lambda := 1$ everywhere and replace $Z^e$ by $Z$ for every occurring module $Z$.

**Problem.** Is this last conclusion also true if $K$ is spherically complete?

## REFERENCES

[1] S. Caenepeel and W.H. Schikhof. *Two elementary proofs of Katsaras' Theorem on p-adic compactoids.* In *Proceedings of the confererence on p-adic analysis (Houthalen, 1986)*, edited by N. De Grande-De Kimpe and L. van Hamme, Vrije Universiteit Brussel, Brussels (1987), 41–44.

[2] A.C.M. van Rooij. *Non-archimedean Functional Analysis.* Marcel Dekker, 1978.

[3] A.C.M. van Rooij and W.H. Schikhof. *Open Problems.* In *p-adic Functional Analysis*, edited by J.M. Bayod, N. De Grande-De Kimpe and J. Martínez-Maurica, Marcel Dekker, New York (1991), 209–219.

[4] W.H. Schikhof. *The continuous linear image of a p-adic compactoid.* Nederl. Akad. Wetensch. Indag. Math. **51** (1989), 119–123.

KATHOLIEKE UNIVERSITEIT
MATHEMATISCH INSTITUUT
TOERNOOIVELD
6525 ED NIJMEGEN, THE NETHERLANDS